Choosy Women and Cheating Men

For Art & Thelma

— Tom.

Choosy Women and Cheating Men

Evolution and Human Behavior

Tom Shellberg

The opinions expressed in this manuscript are solely the opinions of the author and do not represent the opinions or thoughts of the publisher. The author has represented and warranted full ownership and/or legal right to publish all the materials in this book.

Choosy Women and Cheating Men
Evolution and Human Behavior
All Rights Reserved.
Copyright © 2012 Tom Shellberg
v2.0 r1.1

This book may not be reproduced, transmitted, or stored in whole or in part by any means, including graphic, electronic, or mechanical without the express written consent of the publisher except in the case of brief quotations embodied in critical articles and reviews.

Outskirts Press, Inc.
http://www.outskirtspress.com

ISBN: 978-1-4327-8272-6

Outskirts Press and the "OP" logo are trademarks belonging to Outskirts Press, Inc.

PRINTED IN THE UNITED STATES OF AMERICA

Table of Contents

Preface and Acknowledgments i

Chapter 1: Biology and Human Behavior 1

Chapter 2: How to Answer "Why" Questions About Life 15

Chapter 3: The Coolidge Effect and Human Affairs 25

Chapter 4: Why Men Pay For Dates And Women Choose Their Mates (Why men and women are different) 35

Chapter 5: Instinctive Attractions and Aversions 49

Chapter 6: Concealed Estrus, Sexy Bonobos, Porcupine Dildos And Gay Rams (Evolution and the Purposes of Sex) 65

Chapter 7: Tall Bishops and Genuflection Genes 75

Chapter 8: Why Charity Begins at Home (Kin Selection) ... 89

Chapter 9: Vampire Bats, Mother Teresa, and True Love (Reciprocal Altruism) 101

Chapter 10: What's Fun and What's Done 113

Chapter 11: Homo Vestigius 121

Chapter 12: Football, Patriotism, Women, and War (Evolution and own-groupism) 133

Chapter 13: Why Are People Gay?...................................151

Chapter 14: Unintelligent Design 163

Chapter 15: Why Are People Religious?...........................169

Chapter 16: Why We Age and Die
(And why we won't someday)..185

Chapter 17: Evolution In The Third Grade 195

Preface and Acknowledgments

This book was inspired by the wonderful enthusiasm my college students showed for learning in my Evolution and Behavior course. I developed that course because the only introductory behavior courses then (and still) available to most students; Psychology, Sociology, and Anthropology, were woefully limited by the methods and perspectives of social science. I believed that a basic introductory course in the biology of behavior was badly needed to counterbalance the social science courses. So, my course described the basics of Ethology, Behavior Genetics, Sociobiology, and the other biobehavioral sciences. This provided opportunity to learn about the genes and hormones and neurotransmitters and instincts and other biological influences on behavior. And, more importantly perhaps, students could learn for the first time, the natural selection-based explanations, rarely mentioned in social science classes, of why behaviors evolved to be as they are.

Well, I was right to expect students would be interested in this subject matter, but I underestimated how eager they were for this information—the class quickly became one of the most popular courses on campus.

My colleagues in the Biology Department at that time

were not so enthused, to say the least. When I initially proposed the course they voted it down by a secret vote! A secret vote in a science department! Most of them were not teaching any evolution in their general biology courses and they did not think that it was important that students learn about the biology of animal behavior much less human behavior. They did think, however, that it was very important that all beginning biology students should learn all microscopic layers of leaves and stems and the latin names of local plants, and hours worth of details about mitosis and meiosis and the Krebs cycle which nearly all students said were unimportant to them and extremely boring.

The students, though, did think that learning about the biology of behavior was very important. Many, on class evaluations, said it was the most interesting, most valuable class they had ever had, and that comparable courses should be required for most college majors. Soon the enrollment swelled to 200 students each semester. Elementary Education majors, Pre-med and Pre-law majors, Criminal Justice, Philosophy, Psychology, and Art majors as well as Biology majors and many others all took the class.

Thank you students for your encouragement. It helped inspire me to write this book.

And thank you to my friends especially wonderful Sylvia Woodburne and all of you who read some of the chapters and made helpful comments; Anne Doran, John Hazard, Judy Nakdimen, Glenn Weisfeld, Esther Frank, Laura and Steve Thurlow, Bill and Lyn Burns, Elaine Watson, Marel Thomas, and especially my wife Mary Lou who, as always, offered the best advice of all.

CHAPTER 1

Biology and Human Behavior

THE STATISTICS ABOUT marital infidelity are difficult to be sure of because many people are unwilling to be honest, even if they are assured their responses will be kept private. However, several research studies have indicated that 40 to 50 percent of married men (in the U.S) have extramarital sex at some during their lives, but only about one third as many women ever cheat. Why is this difference so great?

Why do so many men cheat on their wives? Even politicians who have so much to lose? John Ensign, David Vitter, John Edwards, Bill Clinton, Mark Sanford, Eliot Spitzer, James McGreevey, and Newt Gingrich, to name just a few in recent years. Even evangelical preachers like Jim Bakker and Ted Haggard whose careers are devastated when their hypocrisy is revealed.

We know from many studies that when men cheat its mostly (80%) for the sex. When women do it's mostly (80%) for reasons *other* than sex.

Why do men cheat far more often than women do, and why this big gender difference in *reasons* for extramarital affairs?

◄ CHOOSY WOMEN AND CHEATING MEN

Gay men have, on average, at least 10 or 20 times as many sex partners in a lifetime as straight men do, but gay women, just like straight women, have far fewer partners.

Why is this so?

The answer to each of these questions requires an understanding of evolutionary psychology. Yes, evolution! It is not enough to know about genes and hormones and neural circuits and stimuli and other proximate causes in order to understand why we behave as we do. To answer most 'why' questions about common human behaviors, we need to refer to what natural selection favored during our evolutionary past.

Cheating fundamentalist preachers will, of course, not be pleased to hear that their sexual escapades have anything to do with evolution, but there is no scientific doubt about it— natural selection has inclined men's brains (and the brains of virtually all male animals) to be relatively promiscuous, and to want sex with *new* partners. Women's brains, in contrast, have been shaped by selection to want committed relationships, and women, unlike men, have little interest in having sex just because the partner is new. The evolutionary reasons for these male-female differences are explained in the chapter on *The Coolidge Effect and Human Affairs.*

Men, of course, are not forced by uncontrollable instincts to cheat. They *can* use better judgment. They don't have to try to seduce the attractive woman they meet at the conference. Nobody is making them flirt with the waitress at the hotel bar. They are not absolutely compelled by hormones to pay for a prostitute. Men *do* have some free will and they *can* make other choices. They could take a cold shower and ignore their biological inclinations. Instead of looking for sex outside their marriages. they *could* go to church and pray instead, (and resolve not to lustfully look at the women in the pew in front

BIOLOGY AND HUMAN BEHAVIOR

of them). This should be an easy choice for evangelical ministers. Biology isn't certain destiny. Men aren't *forced* to scratch when they itch.

But men often *do* cheat even though it is socially disapproved, and even though it is often destructive to their marriages and careers because it is biologically *natural* for men to want sex with more partners than their wives. Some may say cancer is natural too. I am not justifying sexual cheating. The purpose of this book is to describe our natural evolved psychology, not to make moral judgments.

It may appear, so far, that this book is going to be about sexual behaviors, but sex is only one of many topics discussed. **This book is more generally about how evolution has shaped our biological predispositions and inclinations and instincts which evolved many thousands of generations ago, many of which we have shared with our animal relatives for millions of years.**

It's not just questions about sex; if you wish to understand why most people, men especially, like football and other team sports, or why we are patriotic, or why warfare has been so constant it is necessary to understand how **natural selection** has shaped human behavior!

If you want to understand why tall men are preferred as dates and mates, why they make more money, and why the taller candidates for public office usually win, it is necessary to understand **human evolution!**

To explain we think some people are good looking or sexy, or why children play so much, or why men usually pay for dates, or why people are religious; if you want good scientific answers to why most all common human behaviors are the way they are, the *only* way to get much insight is by referring to **evolution by natural selection.**

This book is *not* about our learned behaviors which differ from one individual and one society to another, but rather it is about our evolved genetically-influenced behaviors that are much the same in all societies—behaviors which are human universals—the same in African farmers, Swedish scientists, Chinese cab drivers and British royalty.

The Dark Ages of Behavior Study

Even though the behaviors described in this book are typical of people in all human societies, the evolutionary explanations of many of them will be unfamiliar and surprising to many readers. This is especially true of readers whose academic training in human behavior was limited to typical courses in psychology, sociology, and anthropology taught from the 1940's until at least the 1980's. Biological explanations of human behavior were largely ignored or denied during those dark ages; ever since James Watson and B.F. Skinner founded behavioral psychology, and ever since the environmentally-deterministic Standard Social Science model came to dominate the study of human behavior.

We were taught then that except for some behaviorally uninteresting reflexes, most human behavior is due to learning and acculturation. Boys and girls behave differently, we were told, because of the way they are raised. Boys are given trucks and footballs and girls are given dolls, and thus children are *taught* to behave in gender-appropriate ways. Some people still believe this; social constructionists, gender feminists, behavioral psychologists and many sociologists and Marxists among them.

Sometimes people hold these environmentally deterministic

BIOLOGY AND HUMAN BEHAVIOR

views because they are not aware that boys in *all* cultures behave differently from girls in many instinctive ways regardless of how they are raised.

Another common reason why some people object to biological explanations of human behavior is because their beliefs are influenced less by scientific data than by social and political attitudes. Many Marxists, for example, cannot accept that humans are innately selfish because then a communistic good-of-the-species political system like they espouse would rub against the grain of human nature and therefore would not work unless people were denied the liberty to behave naturally. Likewise some feminists cannot accept that males and females have innate behavioral differences because that would seem to suggest (it doesn't though) that males and females should not have equal rights and opportunities.

We were told that boys in all cultures are more physically aggressive and fight more with one another because that's how society shapes them. But why did anyone ever think this makes sense? Why would all those hundreds of societies all teach boys to behave this way? And how to explain why it is that males of virtually *all* species fight more with one another than females do? Were we supposed to believe that beetles and lizards and mountain sheep all teach their males to physically fight with one another?

Basic male-female differences in humans are very much the same as they are in our animal relatives. Why did so few social scientists fail to realize the significance of this fact? Part of the reason they didn't is because animal behavior was rarely taught even in biology courses, and evolutionary perspectives were rarely applied to understanding human behavior. We didn't realize the significance of the fact that we *are* animals after all.

CHOOSY WOMEN AND CHEATING MEN

If a boy turned out to be gay there was rarely any thought that the reasons might be biological. No, it was assumed that the cause had to be environmental. Maybe his father was distant or his mother was overprotective, or he didn't resolve the Oedipal complex (Freudian nonsense, we know now) which said that little boys go through a stage where they are hostile, even hateful, to their fathers because they are jealous that their father possesses their mother sexually. If you told your psychology professor that you never felt that way about your father and you were certainly not sexually interested in your mother when you were a small child, he was likely to say that's because you *repressed* it (more Freudian nonsense).

Because the study of human behavior largely denied biological influences and was founded on the ridiculous assumption that human behavior is almost entirely the result of experience, a lot of unscientific nonsense (like most Freudian theory) was taught in those years. Margaret Mead, perhaps the most well known social scientist of the fifties and sixties—the Grand Dame of Anthropology—quoted frequently by the Reader's Digest, argued in her very popular books that there were some societies where males and females did not behave differently, and in some societies there was no significant aggression, nor rape or murder, for example. Poor Margaret was not a very good researcher and was duped by the subjects she interviewed. Later research showed that the societies she studied turned out to have the very same male-female differences as all, and the rates of rape and murder in the societies she said were so gentle were sometimes much higher than average. Margaret Mead was a student of Franz Boas, the founder of Anthropology. Had she also studied under a mentor who knew something about the biology of behavior,

BIOLOGY AND HUMAN BEHAVIOR

she might have questioned the paradigm of cultural determinism which dominated those decades and saved her reputation from later disgrace.

If one wondered why some people are smarter than others, or more shy or depressed or better at math, there was little or no mention then of genes or hormones or other biological causes. Learning was everything. We were blank slates at birth and culture shaped all our behaviors, so they said. Steven Pinker has written a rich, fascinating, and comprehensive discussion of why these unscientific ideas persisted for so long and still do today in his excellent book *The Blank Slate* (2002). If one asked during those decades why some people are perceived to be more beautiful or sexy than others, one received the standard social science answer that standards of beauty and sexual appeal are strictly personal (beauty is in the eye of the beholder), or are shaped by culture. It was decades before scientific research proved that our perceptions of beauty and sexiness are not so much personal and culturally determined as we had thought, but instead are mostly universal and are largely determined by instincts shaped during our evolution.

If you asked then why we play team sports or why we are patriotic or why human warfare has been so constant, hardly anyone suggested that this must have something to do with our biology since almost all social animals are altruistic to their own groups and hostile to other groups of their species. At that time only a few ethologists like Lorenz and Tinbergen, most of them in Europe, were studying the biology of behavior, and there was little realization that to understand human behavior one must first understand the general principles of animal behavior and evolution.

If you studied human behavior in the U.S. from the 1940's until at least the 1980's you would have thought we dropped out of the sky, unconnected with our biological past or that we had transcended our animal biology, no longer animals except structurally. There was not much reference to genes or hormones or evolved brain circuits or instincts. Many social scientists even insisted that there was no such thing as human instincts. Some still do! Many Americans still believe the earth was created less than 10.000 years ago!

Invisible Animals

Most of us today realize vaguely that we are animals structurally but few of us seem to realize that we are animals behaviorally as well. That's not just because the social sciences denied for so long that biology had much to do with human behavior but it is also because most biology courses failed to include animal behavior, and schools rarely provided courses in the biology of human behavior. And as Pinker's *The Blank Slate* points out, there are many other reasons why we have failed to realize our human animal behaviors. Again, read his book if you want to understand those dark ages and why biological explanations of our behavior are rejected even today.

In the general biology courses most of us took, we dissected frogs or pigs and learned about animal and human *anatomy* but curiously not about animal behavior much less human *behavior*. There were a hundred courses where psychology and education and pre-law majors were required to learn about the structure of roots and stems and the detailed stages of meiosis for every one which included a unit on the biology of behavior. Those students learned whether the endodermis was to the inside or the outside of the pericycle in the Ranunculus root, and they all were taught the boring chemical details of the Krebs cycle, but almost none of us then were learning why males and females have evolved to behave differently, or why humans and other social animals are antagonistic to other groups within their own species. Biology teachers, like most everyone else, believed that human behavior was all due to learning or was not a suitable topic for general biology courses. In any case, **my intent in this book is to describe many of these evolved human behaviors which we didn't learn about then; behaviors which are the same in all cultures, and which usually are very similar to those of our animal relatives because of shared genes.**

Our emotional responses, for example, are *all* instinctive. Anger, fear, love, joy, sadness, disgust, shame, and many others we rarely realize or talk about. We don't learn *any* of these! We are automatically prewired by our evolutionary past to feel these emotions in response to certain stimuli.

Our facial expressions in showing these emotions are also automatic and inborn and the same in every culture in the world. Even children born blind, smile when they are happy or amused, and frown to show displeasure, and they show the same facial expressions of fear and surprise and disgust and anger as sighted children do. *Our emotions and the ways we express them can sometimes be partly controlled but they are essentially instinctive.*

This book will discuss some of these natural emotional responses, and also our instinctive attractions and aversions, our instinctive sexual behaviors, our biologically-based male and female behavior differences, our programmed learning, our natural aggressive and territorial inclinations, and much more. Everything from why men have "wandering eyes" (chapter 3) to why we are religious (chapter 15).

Some of our biological behaviors, such as our instinctive program to learn language according to innate rules of grammar and syntax, and our pleasurable response to rhythm and music are unique to humans. But most of our biological behaviors are shared with our closest animal relatives. How could it be otherwise? We *are* animals after all. It's not just our anatomy and physiology which are much like that of our animal relatives. Our behaviors are highly similar too because of shared genetic descent.

Some of our basic male-female differences, for example, are almost universal; typical of nearly all animals. Female crickets and frogs, for example, like female humans, are much more picky than male crickets and frogs when it comes to whom they will have sex with, while males of virtually all species, like human males, are biologically inclined to be much more promiscuous.

It's easy to see we are vertebrates structurally, since we have backbones, hormones (like adrenalin and testosterone), a chambered heart, a brain with a cerebrum, cerebellum, and medulla, and a hollow dorsal nerve cord, just like our other vertebrate relatives (fish, amphibians, reptiles, birds, and mammals). It's not so easy, though, for us to realize that we *behave* like vertebrates too. But how could it be otherwise? Yes, our brain is proportionately much bigger than that of any other vertebrate, and we are capable of far more learning, but, after all, our brain is a *vertebrate* brain not a blank slate. Since we *are* vertebrates shouldn't we expect that we would behave like vertebrates? Scientists from another planet would know we are vertebrates even if they couldn't see our structures. They would be able to tell by our behaviors. Our territorial and aggressive, and sexual behaviors, for example, are just as distinctly vertebrate in pattern as our structures are.

BIOLOGY AND HUMAN BEHAVIOR

Among the vertebrates, our closest relatives are the other mammals. Like them, we have hair, mammary glands, a placenta, sweat glands, a diaphragm, an extra-large cerebrum, and a limbic system which produces emotions. It's no wonder since we are mammals that we show many typical mammal behaviors from play to maternal care, and the full range of mammal emotions from jealousy to bonding/love. Why didn't we realize, for example, that since human mothers, just like all other mammal mothers, nurse and love and care for their young that this just might have something to do with the fact that we are mammals?

Our closest mammal relatives, of course, are the other primates, especially, the apes. Like them, in addition to our shared vertebrate and mammal characteristics, we have binocular-color vision, flat nails, opposable thumbs and many other physical traits due to our close genetic relationship. Our very closest primate relatives are the chimpanzees and bonobos with whom we share more than 98% of our genes, so, of course, we share many of the same behaviors from kissing and hugging to political manipulation and deliberate deception. Franz De Wall's books, especially *Chimpanzee Politics, Bonobo,* and *Good Natured* are wonderful sources for reading about the many behaviors we share with these apes.

Proximate Causes of Behavior

Because this book is mostly about the biology of human behavior, there will be some references to genes, since genes play a major, albeit indirect, role in producing all behaviors. Genes also greatly influence our temperaments and abilities, **even our social and political attitudes** as many behavior-genetics studies have shown. Genes influence whether we are shy or bold, inclined toward leadership or not, extroverted or introverted, gullible or skeptical, dominating or submissive, etc. Almost all personality traits and abilities; logic, orderliness, creativity, stubbornness, sociability, fearfulness, etc., are affected by which genes we have. Even the inclination to be religious or non-religious as an adult is significantly affected by genes.

Nobody thinks there is a single gene for being shy or extroverted or sociopathic or religious. Our biological inclinations and instincts result from a combination of many genes. And, it is true, of course, that genes cannot cause any trait all by themselves. All genetic expression depends upon the environment, so even identical twins differ in some ways because small environmental differences cause genes to be expressed or suppressed. Nonetheless, the extent to which genes usually influence human behavior has startled even many behavior geneticists. For more information about human behavior-genetics, one of the best sources is the extensive research which has been done for decades by Thomas Bouchard and his team at the University of Minnesota.

The biology of behavior also has a lot to do with hormones and neurotransmitters and brain circuits produced by the genes. Testosterone, for example, increases confidence and aggressiveness, and is necessary for libido in both sexes. Also, the level present in the fetus during the second month of pregnancy determines, by programming the hypothalamus, whether the child, regardless of anatomical sex, will have a masculine or feminine brain after birth. Girls whose brains receive high levels of testosterone during this critical period will usually show typical male-type (tomboy) behaviors and sexual interest in females. Boys whose fetal hypothalamuses receive too little testosterone grow up feeling like females and

BIOLOGY AND HUMAN BEHAVIOR

usually show female-type behaviors and as adults they are sexually attracted to men.

Near these hypothalamic brain centers which produce gender behaviors and sexual preference there is another center for "falling in love". When patients have had the circuits cut here during surgery they can still be friends, and can still have sex but they can no longer experience the biological addiction we call "falling in love".

Genes and hormones, neurotransmitters, brain circuits, and instinctive responses are, all important in explaining human behaviors. They tell us all about the immediate preceding physical causes or influences which produce our behaviors. But they don't tell us anything about *why* we have the behaviors we do. But that's what this book is mostly about. **The main focus of this book is not on the proximate physical causes like genes and hormones, it is rather upon the evolutionary 'why' questions about our common biological behaviors** (explained in the next chapter). Among the questions explored in this book are the following:

- Why do men and virtually all male animals have such a "wandering eye" interest in the possibility of sex with new sexual partners? (Ch.3)
- Why do humans in all cultures show the same basic male-female differences as occur in other animals? (ch.4)
- Why do we have so much non-procreative sex, and what are the natural evolved purposes of sex? (Ch.6)
- Why do we play as do other mammals? Why don't most animals play? Why do youngsters play the most and why do we tend to get neophobic as we get older? (Ch.10)
- Why such interest in team sports especially by men,

why are we patriotic, and why are we so easily inclined to go to war? (Ch.12)
- Why do we give tall men so many advantages, and why do we lower our heads and bodies when we feel deferential or when we are appeasing? (Ch.7)
- Why do we instinctively like sweets and salt and fat, but not bitter or rotting foods? Why do we think some people are beautiful. and why do we think certain body features are sexy? (Ch.5)
- Why are we usually more altruistic to our relatives and why can altruism often be better described as selfishness? (Chs. 8 & 9)

These questions have been explored by sociobiologists and evolutionary psychologists and other scientists who study the biology of human behavior and their answers (often surprising and unfamiliar to non-scientists) will be described in following chapters as will many other "why" questions about human behavior. I will also discuss some "why" questions to which the answers are more controversial and scientists are not all agreed such as:
- Why are most people religious? (Ch. 15)
- Why do we age and therefore die? (Ch.16)

CHAPTER 2

How to Answer "Why" Questions About Life

IMAGINE FOR A moment you are taking a college biology class, and on the very first day the instructor asks all of the students to try (in a sentence or two) to answer the following questions.
 1. Why do we sleep?
 2. Why do animals (humans too) avoid incest?
 3. Why do some people have light skin while others have dark skin?
 4. Why do women undergo menopause?
 5. Why do humans have far more sex than most animals?

 I have asked these questions of hundreds of students on the first day of class. The answers they gave were remarkably similar every semester regardless of the sophistication of the students (who ranged in age from 18 to 60). If you were in that class how would you answer these questions?
 Virtually all of the students answered the first two questions by saying we sleep to be rested or rejuvenated, and most

said animals avoid incest because they would have defective offspring.

Although it's true that we feel rested or rejuvenated after sleeping, and it's true that mating with close relatives often produces defective offspring, that's not how most evolutionary scientists would answer these two questions. For one thing, scientists say the reason why sleep evolved has little or nothing to do with rest or rejuvenation, and, of course, animals cannot possibly know that if they mate with close relatives their offspring are more likely to be defective.

But there is another reason why biological scientists and philosophers would not give responses like these. These answers are **teleological.** Being rested or rejuvenated is, at best, the *effect* of sleeping, and not having defective offspring is the *effect* of avoiding incest. So, how can these be the causes? When we ask "why" aren't we looking for causes? Teleological thinking confuses cause with effect and makes it seem that the effect or supposed 'purpose' is the cause. If we are to understand why something biological is the way it is we need to identify *causes,* rather than tell effects or supposed 'purposes'.

When the students realized that they didn't give proper scientific answers to these questions—that they told effects rather than causes— they were usually surprised.

If, like the students, you gave teleological answers to these first two questions, you should not be concerned. Most people do! Sociobiologist E. O. Wilson of Harvard and many evolutionary psychologists have suggested we are biologically predisposed to think teleologically. In other words, teleological thinking is quite natural to us even though it is not scientific! Even biology teachers sometimes give teleological answers. I remember one instructor, who, when asked in a hiring interview, why ecologic succession occurs, said "To get to the next stage." We didn't hire him.

Even Konrad Lorenz, the great ethologist who won the Nobel Prize for his studies of animal behavior, was thinking teleologically all those years when he gave "good of the species" answers to 'why' questions about animal behavior. He said that animals restrained aggression toward their rivals and practiced 'birth control' when populations were too dense "for the good of the species."

Incidentally, animals *never* behave for the good of the species, but even if they did, the benefit to the species would be, at best, the *result* of these behaviors, not the cause. Lorenz shouldn't be criticized, though, for thinking that animals behaved for the good of the group. Everybody used to think that way. It was not until the 1960's when understanding of natural selection became gene-focused that this thinking was rejected by evolutionary scientists. And, it wasn't until the late 70's after the publication of Dawkins' *The Selfish Gene* that most biologists realized why good-of-the-species thinking doesn't make evolutionary sense. Still, it is unfortunate that Lorenz didn't realize the teleological trap which interfered with his understanding of why animals *really* behave as they do. If he had recognized this common error of logical thought he would have realized that his for-the-good-of-the-species answers to why animals rarely kill their rivals, and why they control the number of their offspring, didn't explain anything at all.

After a brief discussion of teleology in those classes I told the students we can do better. We can avoid teleological answers. So we started over for a second round. I asked the students to suggest some *causes* this time, no effects or purposes please. They caught on right away, and in this second round they identified all sorts of preceding causes. They realized, for example, there must be chemicals and brain circuits and environmental clues which stimulate sleep. With respect to skin color they said that dark skin is caused by more melanin, and that people with light skin have less melanin. Some of the students even correctly figured out there must be some instinctive mechanism which stops animals from mating with close relatives. We had made progress. The students were listing causes rather than telling effects!

The forth question about menopause is a good one for showing *chains* of causation. Many students answered that the reason women experience menopause is that the ovaries stop producing eggs and hormones. So I said "That's correct. But is everyone satisfied with this answer?" Invariably some students then said "No, because we need to explain why the ovaries stop functioning." So I told them "It's because the pituitary gland stops stimulating the ovaries. "Ok, are we done with this? Do you now understand why women have menopause?" They realized this just begs the next question. We need to answer why the pituitary stops stimulating the ovaries, and so on. Biological phenomena result from chains of causation. We need to go down a ladder of many preceding physical causes all the way to the genes at the bottom rung if we are to understand menopause or skin color or sleep or any biological trait.

Proximate Answers

After this second round of answers, I complimented the students for properly giving causes rather than effects, and explained that the answers they gave this time around are all de-

HOW TO ANSWER "WHY" QUESTIONS ABOUT LIFE

scribed by scientists as **proximate** answers. Proximate answers identify all the preceding physical causes and mechanisms which produce a trait. They are hormones, neurotransmitters, muscles, developmental processes and stimuli, etc. Melanin is part of the proximate explanation of skin color, brain circuits are part of the proximate explanation of what causes sleep, the failure of the pituitary to stimulate the ovaries is part of the proximate explanation of menopause. I stressed that proximate answers, though, correct, tell us *what* and *how* and *when* and *where*, but not **"why"** though. We still hadn't answered these 'why' questions!

What does it mean to ask 'why' questions about common biological traits? What sort of answer should we expect when we ask why some people have dark skin but others are light, or why women have menopause but men stay fertile until old age or why we have fevers when we are sick, or why men are so stimulated by the prospect of sex with new partners (discussed in the next chapter)? When we ask why a prevalent genetically-influenced trait is the way it is, rather than some other way, aren't we asking a very different sort of question requiring a different type of answer than when we ask 'what' or' how' or 'when'? **To evolutionary scientists, answering 'why' questions like these requires that we get past proximate causation to explain in a more ultimate sense why these traits evolved by natural selection.**

So in class, we started over again for a third round and attempted to answer why these things came to be by natural selection, and to do this we need to explain why they are the way they are, *rather than some other way*. To explain why women have menopause it's necessary to be able to explain *why men, and females of most other species, do not*. We need to explain why it's males *but not females* who show the

Coolidge Effect interest in new sex partners. To understand what sleep is all about we must be able to explain why some animals sleep most of the time *but some sleep very little or not at all*. To understand **why** sleep or menopause or skin color or wandering eyes or any biological trait evolved we need to get past proximate answers to figure out why the genes for these traits came to prevail here but not there. The only way to do this is to apply our understanding of natural selection.

Darwin's Master Key

Lamark and some other scientists earlier than Darwin realized that evolution must have occurred. But Darwin (and Wallace, independently) were the first to scientifically demonstrate the fact of evolution more than 150 years ago when they showed that all life on this planet arose and developed by a natural process. Many scientists and historians and philosophers of science hailed this as the greatest scientific discovery of all time. And, of course, they still do—the scientific and social and practical and philosophical implications of the realization of evolution are more awesome and portentous than that of any discovery before or since.

Perhaps even more important, though, than his discovery of evolution was Darwin's theory of natural selection. Natural selection was the master key which allowed scientists, for the first time in human history, to answer 'why' questions about life!

Before Darwin in 1859, biological scientists were limited to proximate explanations. There was no way to give scientific answers to 'why' questions about biological traits. No way to ultimately answer why biological traits are this way

and not that. No way to answer why men are generally bigger than women, no way to answer why males behave differently than females. No way to really understand why species curtail reproduction when density is high. And, no way to answer most 'why' questions about instinctive human behaviors. And, because culture is on a genetic leash as E.O. Wilson has said, without reference to natural selection we cannot even understand 'cultural' phenomena such as standards of beauty (chapter 5), team sports (Chapter 12), why the taller candidates for public office have a big advantage (Chapter 7), or why we are religious (Chapter 15**). Darwin's great contribution to science and philosophy was that he showed us how to answer why questions about life.**

Back to the 'why' questions I posed at the beginning of those classes: having been forewarned about teleological responses which are unscientific, and proximate answers which really don't answer 'why', how did the students respond in the third round of this Socratic experiment? Well, they struggled a bit, unfamiliar with evolutionary answers. Skin color was the easiest for them to hypothesize about. They realized that genes for dark skin must have conferred some advantage close to the equator where sunlight is intense all year, and that genes for light skin must have been favored where there is much less sun. In fact dark skin protects against skin cancer and excessive production of vitamin D, and light skin is advantageous in helping to synthesize vitamin D where people lacking sufficient sun would otherwise not get enough. No wonder there is a gradual increase in skin darkness from Scandinavia to central Africa; Swedes, Finns, and Norwegians being the lightest, Spaniards, Italians, and Arabs somewhat darker, and equatorial Africans the darkest of all. The students, of course, realized that this evolutionary explanation of why some people have light skin and some have dark is categorically different from the proximate identification of melanin, which merely tells us 'what'. For the first time ever most of them now understood **why**!

To understand the leading theory of why menopause evolved it is necessary to understand inclusive fitness theory (kin selection) but since that hadn't yet been presented in class, none of the students could make any good guesses, so I had to explain enough about the grandmother-helper theory so they would get the general idea. This theory proposes that women who historically became sterile at around age 50 left more descendants than those who remained fertile! How could that be? Well, historically, women didn't live nearly as long as now and few would have lived long enough and been strong enough to raise another child after 50. The odds of being able to raise another set of genes at that age would have been very poor. So those women who became sterile around that age but continued to help care for their genes already out there in their children and grandchildren, actually had more surviving descendants—passed on more of their genes. Therefore genes for menopause were favored by natural selection over genes for remaining fertile.

Whether or not this grandmother-helper theory is the best explanation of why menopause evolved didn't matter that early in the class. The students at least were able to understand why natural selection could favor sterility in women but not in men, and that this evolutionary explanation at least attempts to answer 'why', while the proximate answers regarding ovaries and hormones and the pituitary gland only tells what and how and when. Once the students realized that menopause could only evolve in a species where the mother lives long enough to provide extended care for children and even grandchildren, a few of them correctly surmised that we might expect to find menopause in female elephants. Amazing! In one class period the first week of the semester, some students understood enough about the nature of evolutionary explanation that they were already making good scientific predictions.

There usually wasn't enough time in that class session to get to question # 5. So I took a few minutes from the start of the next period to tell the students that humans have sex far more often than most animals and mostly at times when the female cannot get pregnant, and I asked them why they thought this is so. Instead of fishing for proximate answers most of them answered that there must have been some adaptive reason why natural selection favored humans having far more sex than most other animals and they figured out that it likely evolved for some other reason than making babies! (This topic is discussed in chapter 6). Again, after just one class period they were thinking about evolutionary explanations; they were learning how to answer **'why'** questions about life!

I realize that I haven't yet explained the evolutionary reasons why humans and most other animals avoid incest, but I will do so later in the book. With respect to all the other topics about our behavior discussed in this book, there will be some reference to the who, what, where, when—proximate answers, the genes and hormones and brain circuits and neurotransmitters, etc., but the main focus will be on the evolutionary explanations based on natural selection.

In my college classes there were usually some fundamentalist Christians and Muslims but I almost never got any objections about evolution, though I encouraged discussion and never criticized student's opinions. I think that was because

CHOOSY WOMEN AND CHEATING MEN

I *showed* students right away the great explanatory power of selection theory for understanding 'why' questions about life. Instead of objecting they would typically say "Wow" or "Cool" when evolutionary explanations were given. For most of them it was the first time they had ever realized that there is a big difference between proximate and ultimate answers to 'why' questions and that using Darwin's master key allowed them for the first time in their lives to understand the meanings of life.

By the end of the semester I think all of them understood what the great geneticist Dobzhansky meant when he said "Nothing in biology makes sense except in the light of evolution."

CHAPTER 3

The Coolidge Effect and Human Affairs

THERE ARE STILL some small farms where a bull and cow take a fancy to one another and stroll off in the meadow for an ancient bovine tryst. But most cattle breeding today is a very unromantic business. For decades now, cattle have been scientifically bred to produce stock with the best genes for milk and meat production, rapid growth, and disease resistance. Today only a select few genetically-superior stud bulls, each insured for up to a million dollars, father almost all the calves produced for the beef and milk industries. Some stud bulls have fathered over 100,000 calves. There is no more romance left to chance. In fact since most bulls don't measure up genetically, they are never allowed to breed even once.

Because their sperm is so valuable, the stud bulls are hooked up to an artificial vagina and given mechanical stimulation and their semen is then collected to inseminate the cows. To get the maximum amount of sperm these bulls also need some natural stimulation from real cows, so one technique which has been used is to introduce a 'teaser" cow nearby, to help the bulls get aroused. This works well but once the bull has ejaculated two or three times, he loses interest and seems exhausted and incapable of producing another load unless he is given a long rest (refractory) period. That is, if he is presented with the *same* teaser cow. But if a *new* cow is introduced, he springs to life and can ejaculate almost immediately. As long as new teaser cows keep being introduced, the bull will keep ejaculating load after load until he is so physically exhausted his health is threatened so the breeder has to call a halt to let him recover.

This male biological instinct to lose interest in an individual already mated with, and to be sexually stimulated by the prospect of sex with new partners is called the Coolidge Effect, and is well known to animal breeders, zookeepers, and students of human sexuality. The Coolidge Effect has evolved in males because once a female has been impregnated, the male cannot produce any more offspring by continuing to mate with her. (Wild bulls and rams usually refuse to mate with a female more than once!) Males, though, who are genetically inclined to lose interest after mating and to be much more sexually attracted to *new* females, leave far more offspring, so genes in males for attraction to new females spread rapidly by natural selection. It's no wonder the Coolidge Effect has evolved to be a behavioral characteristic of males of most all mammal species and many other animals. The neurochemistry involved includes the feel-good neurotransmitter, dopamine, in the limbic system of the brain. Dopamine level drops after orgasm but rises in anticipation of sex with a new female.

Female humans and other animals do not show the Coolidge Effect because female reproductive potential is far more limited. Even if the female is not already pregnant, switching to a new sexual partner, just because he is new does not produce more offspring, so genes for the Coolidge

THE COOLIDGE EFFECT AND HUMAN AFFAIRS

Effect expressed in female bodies can not be favored by natural selection. It's true that females of many species, even monogamous ones, sometimes abandon their mate for another one or 'cheat' on the side to get better genes or resources for their offspring, but they are not biologically inclined, like males are, to seek sex with a new partner just because that partner is new.

How does this help us understand actual human male and female sexual behaviors? We humans are strongly influenced by cultural expectations and moral constraints but, nonetheless, it is clear from many anthropological and psychological studies that human males show the Coolidge Effect and females do not. Schmitt et al (2003) in a study of over 16,000 people in 52 countries demonstrated that men in every country indicated they ideally wanted many more sexual partners than women did.

Also, it has been established by many scientific studies that men initiate sex much more often with new girlfriends and wives than they do after many months or years with the same woman. In fact, a man's interest in sex with his mate often declines so much that, even if he is young with a strong libido, once or twice a day typically becomes once or twice a week or month, so that she often now wants sex with him more frequently than he does. As his sexual interest in her wanes, she sometimes feels that he no longer loves her as much, or he is working too hard, or is too stressed, or she may suspect he is having secret sex with another lover. Why else would he not want to have sex with her nearly as often as before? Not many wives know about the Coolidge Effect and that it is a natural biological trait of males, so they may worry and may seek counseling.

In most cases the husband loves his wife as much, if not

more, than he always did, and, in any case, men don't even have to like their partner at all, much less be in love to want to have sex. And it's rarely true that a young man or even a 60-year-old has lost his libido. The same man whose sexual interest in his regular partner has diminished, is still very capable of being aroused by new partners, and as evolutionary psychologist Glen Wilson has pointed out, if this man were invited to participate in an orgy, he would eagerly be able to have intercourse with two or three anonymous women in the course of the evening.

The Coolidge Effect influences men much the same the world over so it's not surprising that studies show that males in virtually all cultures have extramarital sex much more often than women do. When women do it's usually because of loveless marriages or dissatisfaction with their relationships or because they get better genes or resources by switching partners, not because they are looking for sex with someone new.

Other evidence of the Coolidge Effect is shown when men and women are asked about their sexual fantasies. **Ellis and Symons (1990) found that men have twice as many sexual fantasies as women, and that men's fantasies and dreams far more often include changing partners, multiple partners and strangers. And, when men are asked how many different sex partners they would ideally like to have in a lifetime they say they would like at least three times as many as women say they would ideally prefer.** In one American study men indicated they would prefer to have an average of 14 different partners while most women said they would like to have only one or two.

Ellis and Symons also asked people who were married or in a relationship, if they had the opportunity to have sex with an anonymous member of the opposite sex, and there was

THE COOLIDGE EFFECT AND HUMAN AFFAIRS

no risk of pregnancy, discovery, or disease, and no chance of forming a more durable relationship, would they do this? Four times as many men as women answered they certainly would do so, while two and a half times more women answered they definitely would not. When the same question was asked of people without steady partners, men answered that they certainly would, six times more often than women did. On a recent *Baggage* TV show where prospective suitors must reveal their 'baggage' before they get the opportunity to be chosen for a date, one man confessed that his baggage was that he subscribed to 10 porn sites. When asked about this he argued that "Everyone needs variety." Of course he was speaking for himself and men in general, not everyone. Women don't need variety per se. Evolution has not shaped them to have the Coolidge Effect.

Fantasies and hypothetical questions aside, when real opportunity arises, except for a few women who are often trying to please their swinger husbands and a few rare 'nympho' women often with very high testosterone, women very rarely have sex with multiple partners or strangers, even if there is little risk of disease or that anyone will find out. Most men, if given such a chance, are usually very eager. In one study by Clark and Hatfield (1989) handsome, genial, young male strangers approached female college students and asked them, after a brief introduction, if they would go to bed with them that night. *None* of the women agreed! But, in the same study, when attractive young female strangers approached male college students, and asked them the same question, *75% of the men agreed! Many of the rest who couldn't arrange for sex that day, asked for a rain check*!

Once one knows about the Coolidge Effect it is much easier to understand why the pornography business is designed almost exclusively for males, and why porn magazines, movies, and websites feature such a large variety of images of naked or near-naked fantasy partners. Even happily married men who would not risk actual sexual relations with a real partner are keenly interested in porn, and many millions of them frequently masturbate to images on the internet. **And, just as predicted by the Coolidge Effect, while men find some images especially sexy and return to them again and again, it usually happens that interest in those images eventually declines and the men find newer images more stimulating. Next month's Playboy is usually more interesting than last month's issue.**

Given that men, like bulls and most mammals, are instinctively attracted to new partners, how many different sex partners do men actually have in a lifetime? Averages are, perhaps, not very meaningful here because the range of variation is so great. Some men, because of low testosterone and therefore low libido, or because of religious beliefs, or lack of opportunity or because they are very monogamously committed, have only one or two or three partners in a lifetime, while other men with a strong libido or lots of opportunities have dozens or hundreds. There is considerable variation between cultures which are very permissive and those with strong puritanical restrictions, and there is great variation from one individual to another but according to an ABC news poll in 2004 the median number of actual different sex partners in a lifetime for American men is around 8 (not counting virtual fantasy porn partners), and for women it is about 3.

This raises an interesting question: If men are promiscuously interested in sex with many partners, why is the average only 8? Why so low? Well, the reason is that most men are having sex with *women*! If a man is very wealthy or powerful

THE COOLIDGE EFFECT AND HUMAN AFFAIRS

and otherwise very appealing he may be able to attract many women for sex, but even then few women will agree to casual sex or one-night stands. Women mostly want sex only in the context of an actual or possible relationship of their choice.

It's quite another story though when it comes to gay men. Gay men show the Coolidge Effect just like straight men, but they are biologically programmed to be attracted to males so they have sex with other gay males. Since *both* individuals show the Coolidge Effect, interest in sex with new partners, the possibilities for lots of sex are almost unlimited. Sure, some gay males don't have sex with anyone, and some stick with one mate, but the majority have many partners during their lives. Many gay men, especially if they are young, or living at home with parents, or married, or otherwise lack good opportunities have frequent sex with strangers and multiple partners in parks and restrooms and backrooms of bars and at the baths. Driven by the Coolidge Effect they may have several partners in one evening without any more introduction than: "Hi, I'm Bob." Often, names are not exchanged at all.

The average number of sexual partners gay men have is, of course, predictably much higher than it is for straight males. The Lambda Report in January 1998 indicated that 40% of gay men, polled at that time, already had had more than 40 sex partners, and Genre magazine in 1996 reported that 25% of the gay men they interviewed had more than 100. In the Journal of Sex Research (1997) Paul Van de Ven et al, analyzing the sexual profiles of 2583 older gay males, reported that the modal range of different partners they had had was between 100 and 500, and that one out of eight older gay males had had between 500 and 1000 sex partners, and another one out of eight had more than 1000. It's not that gay men are hornier or more promiscuous than straight men; they just have much more opportunity since they're having sex with partners who also have the Coolidge Effect. If women showed the Coolidge Effect, straight males would have as many partners as gay men do.

Lesbians, like straight women, do not show the Coolidge Effect, so they very rarely have sex with multiple partners (despite the fictional porn movies made for straight men). Lesbian bars do not have backrooms where anonymous sex takes place, there are no baths for gay women and virtually no porn magazines or internet sites for them. It is extremely rare for any woman gay or straight to have sex with a stranger in a park or restroom.

Finally, why is this phenomenon called the Coolidge Effect? Well, the story is that President and Mrs. Coolidge were visiting a government farm, and she, a bit ahead of him on the tour, noticed that there were at least a hundred hens but only one rooster! So she asked the guide if that one rooster mated with all the hens. The guide said "Yes ma'am, that rooster mates many times every day." Impressed with this vigor, and implying, perhaps, that Calvin was not much of a sexual stud, she told the guide, "You tell that to the President."

When Calvin came by the chickens a bit later, the guide told him that Mrs. Coolidge wanted him to know that the rooster mated many times every day. Calvin reportedly said "Same hen every time?" and the guide said, "No, Mr. President, it's a different hen every time!" whereupon the President said "Well, You tell that to Mrs. Coolidge!"

What should wives and girlfriends do if they discover their man has a stash of porn magazines or cds, or is spending time enjoying porn on the internet? Perhaps the best advice, if you want to keep your relationship strong, is to respectfully look the other way. He's responding to his natural Coolidge

Effect inclinations in a harmless way. As long as he is not having a relationship with anyone else and not threatening your relationship, what's the problem? You might even want to encourage him to enjoy porn, especially if you are not having sex very frequently. It will likely decrease his interest in having sex on the side which could lead to a relationship that could threaten yours. In any case don't assume if he's into porn that he doesn't love you as much as he did before. That's probably not the case. Men who are powerfully in love with their mates like porn just as much as men who don't have any love relationship.

CHAPTER 4

Why Men Pay For Dates And Women Choose Their Mates
(Why men and women are different)

IT'S TRUE THE world over; it's almost always men who pay for dates. We take it for granted, but it could be the other way around. It could, theoretically, be women who usually pay. Why not?

Friends often go Dutch, but if a man wants more than a platonic friendship, it's expected that he should pay. **One thing women usually require of a potential mate is that he is willing to provide resources for her**. If he is just getting started in a career but has good potential, she may wait, but he had better be generous with what little he has, especially, if he wants her to have sex with him. He should definitely pay for the date even if its just a movie and burgers. Why is this so?

Why isn't it usually *women* who pay? To understand this, and many other questions about why men and women behave differently, it's necessary to understand the evolution of the main sex differences between males and females. Just as

CHOOSY WOMEN AND CHEATING MEN

males are physically different from females because of a different evolutionary history, so has natural selection shaped male and female behaviors differently. Many behaviors good for the goose have not been good for the gander and vice-versa.

Ethologists and sociobiologists and evolutionary psychologists who study the biology of behavior have long known that, from insects to humans, many behaviors have evolved to be typically different in males and females. Darwin knew this a century and a half ago. He observed, for example, that males commonly fight over females, but females almost never fight over males, and he saw that while males will eagerly mate with any female, females, are very choosy about whom they will mate with, and they often require the male to provide resources to them before they will mate. It was clear to him that these differences were due to evolution but it was not so clear to him why natural selection would favor these differences. We had to wait for Robert Trivers.

Trivers, a brilliant evolutionary theorist, realized in 1972 that the most essential difference between males and females is the amount of parental investment (PI) each sex provides for the offspring. This is the Rosetta stone, the critical starting point, if one is to understand the different evolutionary paths of male and female behaviors. If you want to understand why women and men behave so differently, you'd be better off to forget traditional psychology and sociology and concentrate on evolutionary biology.

Trivers pointed out that it is almost always the female who provides the greatest PI. Right from the start her investment is much greater because she provides the egg which is thousands of time bigger than the sperm provided by the male. The egg is so much larger because it contains, in addition to half the genes, *all* the nutrients for the developing embryo.

WHY MEN PAY FOR DATES AND WOMEN CHOOSE THEIR MATES

The sperm provides nothing except half the genes. If we use the analogy of making a cake to making a baby, the male and female each provide half the recipe but the female provides *all* the ingredients!

But that's just the beginning of this unequal investment in the offspring. In all those many species where the embryos develop within the female's body, she is also stuck with providing the oven! She must 'bake' the fetus inside her until it finally pops out as an egg or infant. Meanwhile the male can, and usually does, look for other females to sow his seed.

After birth, if there is any parental care at all, it's mom again who is usually stuck with feeding and protecting the youngsters until they can survive on their own. It's true in a few species the dad pitches in, as male birds do, to incubate the egg and help feed the chicks, but birds are quite unusual. Most animal dads do nothing at all. Where was Bambi's father?

In mammals, the unequal investment is even more extreme. The female must carry the fetus for weeks or months because prenatal development is slow. And, after birth, again it's the mother mouse, the mother lion, and the mother deer, the female, usually all by herself, who must protect and feed and teach the young, often for many months or years. In some species, as with lions, the male continues to guard the territory providing a little investment, but not much compared with what the female must invest. It's as though the female invests a thousand dollars and the male invests only one dollar but when the investment pays off for both of them (offspring are produced), they split the profits.

If you're feeling sorry for females at this point, though, hold on. As you will see, females have countered with some impressive evolutionary tricks up their sleeves.

Trivers realized that because males typically invest so little, they will evolve to compete for that investment. The most common way that males compete is by fighting. Most everyone has seen animal behavior films showing male alligators or male elk or male elephants or leopards or lizards or hippos fighting fiercely with other males, threatening, goring, biting, charging, locking horns inflicting injury and sometimes death. Why? It's not just because males are more aggressive or have more testosterone or simply like to fight; it's all about access to sex with females and the parental investment they provide. That's why most of the serious fighting happens during mating season.

Because females invest so much, they have evolved to be very choosy about whom they will pick as fathers. Even if they don't get anything else from the males they pick as mates, they can always go for good genes for their offspring. And they do. Females mate with the winners of the fights. Even some female plants are picky. They don't let just any old pollen grow down their stigmas.

The greatest advance ever in our understanding of why males and females behave differently happened when Robert Trivers realized that the sex with the least parental investment will always evolve to compete for the sex that offers the most, and the sex which has the biggest parental investment will always be the sex that chooses whom to mate with. Proving this point there are a few rare species (like Jacanas) **where the male actually has the largest P.I. and just as Trivers's theory predicts, the females compete over males, defend the territory, and otherwise show male-typical behaviors!**

What's Good For the Goose

Because females have so few opportunities to produce offspring, and invest so much in each one, females have evolved to be extremely careful and discriminating when choosing mates. A mistake, like mating with a male of the wrong species or with a genetically inferior mate who does not supply good genes needed for the survival of her offspring, squanders a precious opportunity to reproduce. In mammals, one small mistake in choosing a mate can cost a year or more of her reproductive life. For males though, there is little or no cost to mating with the wrong species or a genetically inferior female. The only cost of a poor mating choice for males of most species is a few minutes, and a few drops of sperm (which will be quickly replenished). The best general strategy for most males of most species is to be indiscriminate fertilizing machines. Don't hesitate to consider if she is the girl of your dreams. Have sex with every female you can. That's what pays off genetically, that's what natural selection has favored. It's no wonder males so often make mating mistakes, trying to mate with the wrong species or with dummy models which don't look much at all like females. No problem; natural selection has shaped males to be relatively promiscuous and indiscriminate. It's the best way to spread your genes if you don't have much P.I.

So females are the one who choose. How do they choose and what are they looking for? The way that females choose is usually by first 'requiring' the male to do courtship. Did you ever wonder why animals go through all those elaborate courtship rituals, dancing and such, when they could just mate and be done with it?

CHOOSY WOMEN AND CHEATING MEN

From spiders to ducks to people, the main reason why males court (sometimes with highly ritualized displays) is because they have to demonstrate good genetic fitness. Dances and vigorous physical demonstrations are one common way to do this. This allows females a chance to look the male over, check him out, and evaluate him (subconsciously in almost all cases). Females are instinctively tuned in to indicators of good health, strength, good immune systems and other traits which suggest genetic fitness. They, of course, don't 'know' what genes he has: they only have to know what they like. Imagine you are a female bird, and suitor number one has bright feathers, and perfect body symmetry and he does an energetic courtship dance that lasts 15 minutes, never missing a step. Next, suitor number 2 starts to dance but he falls over a couple of times and after 3 minutes he is gasping for breath too exhausted to go on. You notice too that one of his wings is longer than the other, he is quite scabby, his feathers are dull, and he has left a puddle of diarrhea on the dance floor. If it's between the two of them which one do you pick?

You don't have to *know* that vigor and stamina, glossy feathers and good symmetry are all indicators of good genes, and you don't have to *know* that rapid exhaustion, scabs, dull feathers and diarrhea are all signs of genes you shouldn't choose for your chick. You only have to know what you like. Females who choose suitors like number one had far more surviving offspring and passed on more genes, than those who were attracted to males like number 2. That's how natural selection works. Females who just happen to be attracted to indicators of fitness, even though they don't understand why, have more offspring, so in time females have evolved to make better and better choices when it comes to mates.

When males fight with other males during breeding

season, we don't think of this as courtship, but it serves much the same purpose. Females know who the winners are and they usually won't mate with losers, so in a sense females 'require' the males to fight over them. It's in the female's best genetic interest to have the males fight. In species where the males must compete physically, this allows the females to pick the genetic winners. By choosing the big bruiser winners, they have sons who are likely to be big bruisers who will therefore pass on their mother's genes in the next generation.

Anthropologist Irven De Vore has pointed out that **it's as though males are a breeding experiment run by females. Because of female choice, males have been largely designed by females! Women who complain about male behaviors should be aware that it was their great, great grandmothers for thousands of generations back who made men the way they are today!**

Many studies have shown that human females, like females of most species, pick mates with good genes. But what else? What else are women looking for when choosing mates? David Buss, at the University of Texas, and his colleagues have done extensive research on this question. They conducted a worldwide study of mate choice, the biggest study of its type ever done, involving many thousands of people in 37 societies from Finns to Zulus.

In all of the societies they found that women greatly preferred tall, strong, vigorous, athletic men as mates. Not surprising. This is typical of female animals in polygynous species (See the chapter on Tall Bishops and Genuflection Genes) where a few males mate with many females and therefore many males don't mate at all. Our ancestors were quite polygynous and about 80% of human societies still are. Because human females, like females of other polygynous species, tend to prefer males who are strong competitors, able to dominate other males, this has lead to the evolution of behaviors even in young boys which encourage physical competition and status-seeking. Boys fight physically far more than girls do, they brag and boast more, they are much more interested in weapons and conquests in sports or video games, and they engage in much more risky behavior because they are practicing how to intimidate and dominate, and achieve high rank and eventually be chosen by women as mates. And, of course, because men must compete for status, it's no wonder they are much less likely than women to reveal vulnerabilities and more likely to want to be in control. When couples drive together, unless the man has some problem like poor eyesight, it's the man who usually drives. Most men prefer to be in control and most women like them to be. Here's a case where both sexes comfortably agree with what feels natural.

But back to the question of why it's men who pay for dates; **in all of the societies Buss studied, women** not only preferred tall, strong, healthy males with good genes; they also **showed very strong preference for males with good resources and financial prospects.** Given that human females have such extreme parental investment, it's not surprising that natural selection has shaped women to require men to contribute. It's not literally true, of course, that diamonds are a girl's best friend or that women should be characterized as Madonna Material Girls, but **Buss's research showed that women, the world over, valued resources and financial prospect in males twice as much as men valued the same in women. Many studies have shown that most women are especially attracted to wealthy men so much so that they perceive wealthy men even to be better looking than they actually are.**

WHY MEN PAY FOR DATES AND WOMEN CHOOSE THEIR MATES

Because women contribute so much more parental investment in children (and this is especially true in most pre-industrial societies), men are often required to pay a price to the bride's family before they agree to her marriage. Daly and Wilson (1988) studied 680 cultures and found that in 58% of them the groom must pay a dowry to the bride's kin! In only 3% of the societies studied (where there were relatively few wealthy men) did the bride's family pay a price in order to compete.

In many species one of the most important resources females require males to provide is a territory; usually an area which provides shelter and is rich in food resources, is relatively safe from predators, and is defended by the male. Females of territorial species will usually not mate with a male until he has established a defended territory. Female birds often require, as well, that the male provides a nest before they will mate. Males without territories and/or nests do not mate. In fact, in some species such as red-winged blackbirds, females will often share a male who has an especially good territory rather than mate with a male whose territory is poor. Similarly human females in many societies polygynously choose to share a male with other wives when relatively few males have most of the resources. It's better for them than mating with a male of their own who has few resources. Perhaps some feminists will not like this idea, but polygyny evolves where it is the best interest of females. It is, in fact, far more common in primitive human societies than monogamy. As mentioned above, about 80% of all human societies are officially polygamous. This is largely because of female choice.

Women also (far more often than men) choose mates with higher social status. This is partly because men with high social status have more access to resources and provide other advantages for their children.

In every society, Buss also found that women prefer older men (by an average of 3 years). This is usually an indicator of better resources since older men generally have more. Also, older men are typically wiser and more experienced.

Good genes and resources aside, one of the most interesting findings of Buss's study was that women in all of the societies thought it was very important that prospective mates demonstrated love for them, and showed reliable signs of commitment. Lovingness and commitment were also very important to men. Both sexes were quite alike on this measure.

Another trait that women reported was important in their choice of mate had to do with men's attitude toward kids. Women favored men who were affectionate toward children and interacted well with them.

Because of female choice, women have designed men not just to be big and strong and healthy but also to be far more loving, much more faithful to their mates, and much better fathers than males of any other mammal species.

Buss's study also found that both sexes value intelligence, sense of humor, dependability, emotional maturity and several other traits when evaluating prospective mates. These preferences seem to be characteristic only of humans; bears and chipmunks and penguins don't seem to care if their mates are emotionally mature or if they have a good sense of humor.

What's Good For the Gander

While males are far less choosy about whom they will have sex with, especially if it's a one-night stand, men do have strong evolved preferences for specific traits in female sex partners, and what turns on men the most are signs of fertility. It doesn't matter if he's barely out of high school, not ready to get married, much less have kids, or if he's 65, starting retirement and definitely does not want to have any more children, men are most sexually attracted to women with good reproductive potential. Studies of many different cultures all over the world have shown that men are most attracted to women who range in age from the late teens to the late twenties. This is exactly the time when women are most fertile and likely to conceive! Several cross-cultural studies have also shown that men are most attracted to women with a waist-hip ratio of .70. Again, women with this somewhat hourglass shape (slimmer waist, wider hips) have the greatest potential fecundity. Having a fairly slim waist is also, an indicator that the woman is not pregnant, and having wide hips indicates that the woman can probably deliver children. Men, of course, don't think about *why* they find shapely 20-year-old women so much more sexually attractive than extremely skinny or fat 70-year-old women. They're not thinking about natural selection in the singles bar. They don't realize that evolution has shaped them to be attracted to fertile women. Like the female bird who is attracted to vigorous males with glossy feathers or the female elephant seals who swoon over the biggest bruisers, men don't have to understand that their sexual desires are a result of evolution.

Because of this instinctive sexual attraction toward signs of fertility, as men get older, in every culture studied, they prefer women as mates who are increasingly *younger* than they are. Men in their 30's usually prefer women who are about 5 years younger, while men in their 50's commonly prefer mates who are ten to twenty years younger than themselves! It doesn't make much social sense for a 55-year-old man who doesn't want any more kids, to marry a woman a generation younger whose cultural experiences are so different and likely much more modern, and who is likely to live 30 years longer than him. What are they going to talk about? How long is he going to be able to go Salsa dancing on Friday nights? Wouldn't he be better off with a 50-year-old woman who shares more of his life experiences?

You don't have to tell this to the hostess of *The Millionaire Matchmaker* TV program. She's always trying to match up unmarried millionaire men with desirable women of approximately the same age, but almost always, the millionaire men want women who are much younger than themselves. Even men who definitely do not want to have children are usually most attracted to women who physically show evidence of fecundity/ fertility. Nowadays this behavioral preference should probably be considered to be vestigial and often maladaptive.

In many animal species, males show a preference for virgins, and in some species will *only* mate with virgins. You might think, if you didn't know about evolution, that human males, at least, might prefer females with lots of sexual experience, but no, this is definitely not the case! In many human cultures too, men, just like males of many other species, show a strong preference for virgins, so much so that in some societies men will not marry a woman unless it is demonstrated that she is, in fact, a virgin. This is understandable to evolutionary psychologists. If she has had sex with another man she could be pregnant and he could be cuckolded into raising some other man's child. It's very significant here that Mary, the mother of Jesus was said to be a virgin as were the mothers of messiahs in many similar primitive stories long before Christianity. If Mary had actually had sex with her husband,

Joseph, then it would not be clear whether Jesus was the son of God or the son of Joseph.

Men are almost never expected to be virgins when they marry. In fact a woman might have reason to be nervous if he was. There is quite a double standard here, but again, this makes evolutionary sense. A male can have a lot to lose genetically if his mate has had sex with another male; his children may not be his. But it's no skin off a female's genetic-fitness nose if her mate has slept around. Her children are still hers.

This double standard helps explain why the reasons for male and female jealousy are usually very different. Statistics from societies all over the world show that women most often become jealous if their mate gives resources to another woman, or if he shows commitment or love for the other woman and therefore threatens their relationship.

Men, on the other hand, become jealous, sometimes violently so, if their mate has sex with another male, or they even suspect this is happening. **Male sexual jealousy is the main reason worldwide why men batter and kill women.** (See Daly and Wilson's book *Homicide*). A young woman may feel flattered if her boyfriend gets angry and jealous when other guys look lustfully at her, and she may not mind if he's always watching over her, asking where she's been and where she's going. She may feel he is just being protective. But these are dangerous warning signs.

There are several male-female differences which have evolved that are not a direct consequence of differences in parental investment; the Coolidge Effect is one, for example, described in the previous chapter. Another is the greater mathematical and spatial ability (on average) shown by males, especially the ability to visualize 3-D objects rotated in space. (Males with high testosterone and females with relatively high testosterone and or low estrogen are generally better at math and spatial ability than average for their gender). And, males, even as infants show greater interest in objects especially if they are moving These differences may have evolved because of the long history of male hunting, territorial defense, and warfare; chasing, running, aiming, and for navigating far from home. (It's still true today that boys in all cultures generally wander farther from their home base than do girls).

Women, on the other hand, are generally better at reading, do better on vocabulary tests, and score higher on measures of agreeableness, compassion, and empathy, are more socially sensitive, are better at recognizing non-verbal cues, and, as infants, are more interested in faces of people than things. The long history of intense mothering and socializing children was surely part of the reason why these abilities evolved, but it is also likely that women with their greater social sensibilities have played a major role as peacemakers and subtle politicians in maintaining social harmony within the group.

There are other evolved male-female differences in behavior which are not related to parental investment, but **differences in parental investment explain many of the most important behavioral differences of males and females; why males in virtually all species and societies compete and fight physically more with one another than females do, why females are so much more picky when it comes to whom they will mate with, why men have been largely designed by women, and why it's men who pay for dates.**

CHAPTER 5

Instinctive Attractions and Aversions

SUPPOSE YOU WERE lost in the woods and had no food for days, and you came upon some strange berries which taste sweet. Should you eat them? Well, if you use evolutionary logic you probably should. Sweetness says "Eat me." It is a property which has evolved to attract fruit eaters, who by eating the fruit, spread the seeds thus helping the plant reproduce. Liking sweetness for millions of years helped *us* reproduce too. Ripe fruits are sweeter and far more nutritious than those which haven't ripened yet. Fruit-eating animals including our primate ancestors who liked the taste of sugar therefore survived better than those who didn't, so our sweet tooth evolved by natural selection. Of course, as indicated in the chapter on vestiges, our attraction to sweetness has become rather maladaptive now that we have so much non-nutritious sugar today which helps promote tooth decay and obesity and diabetes.

It would be best if you ate only a few of the berries at first, though, and waited several hours before eating more. Just in case they are ok for some animals, but are poisonous to humans. If you became nauseated or had cramps, or other signs of poisoning, you would probably realize it must have been the berries. Even if you didn't, you wouldn't want to eat them again because an instinctive aversion to them would automatically happen. It's called food avoidance conditioning. When an animal eats a new food and gets sick within a few hours afterward, the animal automatically feels aversion to that food in the future. This happens even if it was not the new food which caused the sickness! If a person, for example, has rice pudding for the first time and gets sick a few hours later, not because of the pudding, but because of the flu, they will thereafter, often for many years, be disgusted by even the smell of rice pudding. In this case the aversion is a mistake but usually it prevents us from eating food again which made us sick before.

What if instead of finding berries, you find a plant with leaves which taste extremely bitter. Should you take a chance and eat them? Probably not. Bitter chemicals (often alkaloids) are the plant's way of saying "Don't eat me." Bad tastes are usually a warning that the plant is poisonous. Eating such plants may cause severe pains, convulsions, diarrhea and even death. Bad tastes and poisons have evolved in many plants because they protect the plant from being eaten. Things which taste bad to us usually *are* bad for us. Just as natural selection has shaped our instinctive attractions, it too has shaped our aversions. People who disliked the taste of bitter plants, survived and thus passed on the genes for this aversion far more often than people who liked bitterness.

Another natural instinctive aversion to some foods happens often during pregnancy, especially in the first trimester. This 'morning sickness' has evolved to help protect the developing fetus from common toxins in foods which can cause damage and miscarriage. Embryos and fetuses are far more

INSTINCTIVE ATTRACTIONS AND AVERSIONS

susceptible to these toxins than adults are, so morning sickness evolved because it helps protect the developing child from possible harm. Research has shown that women who do not experience morning sickness are much more likely to miscarry. Better to be sick than childless.

We have natural attractions and aversions to many smells as well as tastes, and the same principle holds true. If it smells good it probably is good for our genes, and if it smells bad it probably is not. If the only thing you could find to eat was the rotting carcass of a raccoon dead for a week you might be tempted at first if you were hungry enough. But as soon as you opened the swollen belly and the noxious gasses produced by decay caused you to gag, this instinctive warning would surely discourage you from having even one bite. Eating rotten flesh is fine for buzzards which have evolved mechanisms to deal with it, but it is very unhealthy for humans and most animals. You might be ok though, eating the maggots in the carcass. It might depend on whether the bacteria and the dangerous poisons were broken down or were still present in the bodies of the maggots.

Spiders, scorpions (and even insects, generally) and snakes cause instinctive aversive feelings and sometimes disgust or fear in many people even though very few of them pose a serious danger in our modern environments. They did pose more of a threat in our ancestral environments though, and we have retained these evolved natural aversions even though they are not so adaptive nowadays. We still freak out when a harmless centipede runs across the bathroom floor or a harmless earwig crawls out of the bookcase. Better to be safe than bitten.

Flowers, on the other hand, are universally considered to be attractive. They look beautiful to us (and to pollinating insects like bees) and they smell good too. This strongly suggests that attraction to flowers was adaptive for us just as it is for bees. Why? Not because we eat flowers very commonly, or use their nectar, but probably because they promise fruit to come (all fruits develop from flowers) and our ancestral diet consisted substantially of fruit. Cats and dogs and other carnivores, though, don't think flowers are attractive, or that they smell good, because attraction to flowers would not be adaptive for them. Because of the realization that our natural common attractions and aversions have been shaped by our evolutionary past, this has led to much research in the promising young science of Darwinian aesthetics.

We also have many instinctive attractions to what *feels* good, and aversions to what feels bad, and the same general rule usually applies: if it feels good it usually *is* good for survival and reproduction, and if it feels bad it is not. Sex has evolved to feel good because it reproduces our genes, and pain feels bad because it threatens their survival. Pleasure says do it again, and pain says do not do that again.

Of course, pleasure and pain feelings are only adaptive for animals capable of memory and learning and modifying their behavior. What would be the point of a tree experiencing pain? Organisms which cannot change their behavior as a result of learning from experience should therefore not be expected to experience pleasure or pain. Mammals, and humans especially, probably feel the most pleasure and pain because they are the most capable of learning and changing their behavior. Does a fish feel the pain of a hook? Who knows? How would we know? Fish have memory and can learn but they can't change their ways nearly as much as mammals and not even remotely as much as we can. One might guess that fish do feel pain but it is not as intense or important to them. Does a worm feel pain from the hook? Again, we don't really know for sure, but since their learning ability is much more limited, if they do feel pain it's probably not as bad as what a fish feels, and not nearly as bad as what we feel.

Beauty and Sex Appeal

Many of our instinctive attractions are visual and they affect our perceptions of what we consider to be beautiful or ugly such as what makes people physically attractive. Research studies have repeatedly shown that attractive people get more attention from the opposite sex, that they tend to get more votes if they run for office, and in general they are higher paid. The also tend to receive more leniency from judges, and attractive children commonly get more attention from their mothers and higher approval from their teachers. Nancy Etcoff's book *The Survival of the Prettiest* (1999) presents a lot of data showing that we instinctively give good-looking people many advantages.

Even babies are innately biased in favor of good looks. Judith Langlois at the University of Texas did research where adults ranked photos of people's faces on attractiveness and then they showed the most and least attractive photos to 6-month-old babies. The babies looked longest at the photos of the most attractive people and it didn't matter what the race or gender or age was of the people in the photos.

One rather surprising finding of research done by Langlois and others is that people rate computer-generated composites of 16 and 32 faces to be much more attractive than any of the individual faces used to make the composite! Her studies and others indicate we instinctively prefer average looks over even slight extremes. If your eyes are a little bit smaller than average, or your jaw is a little bit bigger, or the distance between your nose and your mouth is a little bit greater than average, you will not be thought to be as attractive as a person who is exactly average. Evolutionary psychologists believe the reason why average faces are considered more attractive is because they indicate greater genetic heterogeneity (a greater variety of different genes). It has long been known that hybrids produced by genetically different parents are generally healthier than individuals who have too many of the very same genes. This may also explain why research done by Dr. Michael Lewis at the University of Cardiff found that faces of mixed-race people are considered more attractive than faces of people whose parents were of the same race.

But what else is it, other than being average, that makes someone good looking? What makes a woman beautiful? What makes a man handsome? And what makes anyone, man or woman, sexy? Well, we used to think that depended on whom you asked. We thought that opinions of what makes someone good looking or sexy were strictly subjective and personal, or influenced by one's culture. In other words we thought beauty was strictly in the eye of the beholder or the eyes of different societies. We didn't imagine that there might be objective standards of beauty and sexiness that people in all cultures would agree upon. But that was before scientists began to study the biology of beauty. We know now that beauty is not so much in the eye of the individual beholders; **people all over the world in very different cultures all agree on what makes someone good looking. Irish and Egyptians and Koreans and primitive African tribesmen all agree on what makes a person good looking and also what makes them sexy. This means, of course, that our**

INSTINCTIVE ATTRACTIONS AND AVERSIONS

perceptions of attractiveness in others are not so much a matter of personal taste or cultural influence but are, instead, largely instinctive.

What makes a person beautiful or sexy, other than being average (genetically heterozygous), is generally the same thing that makes a frog or a penguin or a moose look attractive and sexy to prospective mates. It's evidence of good genes! Yes, strange as this may seem at first, there is a great deal of scientific research which shows that the main thing which make animals and humans good looking and sexually appealing is evidence of good genes! If he makes you swoon he has probably got good genes which you should choose for your kids, and if she's really hot, she's genetically fit, reproductively ready and likely to give you healthy kids. We can usually trust our instinctive standards of beauty and sexual attractiveness just as we can trust fruits which taste sweet.

Many specific physical traits indicate genetic fitness which we find instinctively attractive. Women, for example, are attracted to height in men (see Tall Bishops chapter) and to broad shoulders and muscles (if not too extreme) and rugged facial features (prominent brows, large jaws etc., all of which are shaped by testosterone which masculinizes features), but they are not physically attracted to men with small rounded baby-like, or feminized faces which result from estrogen, especially when they are in the most fertile part of their cycle. It may be that women are naturally attracted to height and strength and indicators of good testosterone level because high testosterone males are better protectors of their women and children and they are also more fertile. This attraction to indicators of testosterone may also explain why so many women find men's facial and body hair (chest, arms, legs,) sexy. In any case, most women are instinctively attracted to men who look masculine and are least attracted to men who look feminine.

As mentioned in the chapter on *Why Men Pay for Dates,* men, of course, are especially attracted to *young* women because young women are the most fertile. Women do not favor youth so much in men, though, because men's fertility does not drop substantially with age as it does with women. As indicated, men, according to many studies, are especially attracted to women whose waist-hip ration is 0.7 which suggests not only that they are not pregnant, but also is a good indicator that they are fertile and able to deliver children.

Most men also like rounded rather large female buttocks. This may be because our female primate ancestors, like modern female chimps, showed they were receptive to sex with swollen genitals which look quite like bums from the rear. Women's butts may instinctively stimulate sexual interest in men for the same reason that bum-like sexual swellings in female chimps are instinctively appealing to the males.

Most men find female's breasts to be sexually stimulating too, but there are different theories about why this is so. One theory is that they mimic the buttocks. In fact if just the cleavage is shown in photos most men can't distinguish breasts from bums. Since humans, unlike most other primates, mate mainly face to face, breasts may have evolved to be so much larger in humans than in other primates because they put the long-attractive-to-males swellings right up front. In any case, one study done at Purdue showed that women hitchhikers who added two inches of padding to their bras got twice as many offers of rides.

Most men also like long legs on women. It's likely because women's legs lengthen rapidly under hormonal influence at puberty signaling sexual readiness.

INSTINCTIVE ATTRACTIONS AND AVERSIONS

When a man spots a woman in the bar or the church picnic whom he finds especially sexy he is probably not thinking his instant attraction has anything to do with evolution. And when a woman sees a man in the crowd whom she notices is very handsome, she is probably not aware that her attraction is instinctive. Neither men nor women are often aware that their perceptions of what makes someone good looking or sexually appealing have been shaped by evolution, and what they are really attracted to is genetic fitness and reproductive promise in prospective mates.

What constitutes good genes in a human is rather different, of course, from what amounts to good genes in a moose or a penguin or a parrotfish. But there are some traits that are universally sexy because they usually guarantee good genes. Evidence of good health such as demonstration of a strong immune system and resistance to parasites is one of the most important.

Parasites are a constant threat to all animals. Their threat has been so great for so many millions of years that some scientists believe that sexual reproduction evolved as a countermeasure to foil parasites by changing the genetic combination lock every generation! Parasites sap strength, often disfigure, cause sickness and often death. If an infected animal does not have good defenses against its parasites it will usually act sick, coughing or retching, or be listless without vigor, and it will usually look sick, with scabs or sores, or hair falling out or dull plumage or poor coloration. Animals who mate with individuals who act sick or look sick do not have as many surviving offspring as those who pick healthy mates, so animals have evolved to be attracted to partners whose immune system and other body defenses are strong enough to deal with parasites. Sick animals who cannot overcome the ravages of parasites are not perceived as sexy by prospective mates. What *is* sexy to animals is vigor and stamina, unblemished skin, glossy plumage, good coloration and other signs of resistance to parasites. We humans are too, sexually attracted to signs of good health, and are averse to indicators of sickness. Even the kindliest of humans are instinctively repelled by oozing sores and badly disfigured faces and bodies. Even zits can be a turn off.

Symmetry is another indicator of genetic fitness, and animals in many species, from scorpion flies to barn swallows are most attracted to those potential mates who are the most symmetrical. Why is symmetry so important in picking mates? Well, if one eye is higher than the other or one leg is longer than the other, this is a sign of developmental instability which indicates the individual is not genetically as fit. If one wing is a little shorter than the other, the animal is not as good at catching prey or escaping from predators. Something has gone wrong. Animals who chose the most symmetrical mates had more surviving offspring so the preference for symmetry has commonly evolved in many animals. Females, of course, are more sensitive than males are to picking mates who are parasite-free and highly symmetrical, because females have far fewer reproductive opportunities and invest much more in each offspring.

Several studies of human beings in many diverse cultures have show that we are very much like other animals in our strong preference for symmetrical mates. In every culture people with the most symmetrical faces and bodies are considered the most beautiful. It doesn't matter if one is a Ugandan, an American, a Japanese, or a Mexican: the most symmetrical people are considered the most beautiful. Our bias in favor of symmetry is universal and instinctive. Even

INSTINCTIVE ATTRACTIONS AND AVERSIONS

babies prefer to look at pictures of people with the most symmetrical faces.

Women can even tell by smell which men are the most symmetrical! Randy Thornhill of the University of New Mexico had women sniff undershirts men had worn and slept in, and he found that women strongly preferred the smell of shirts which had been worn by the most symmetrical males! This was true even when the women couldn't consciously smell anything, and were subconsciously detecting pheromones in the sweat.

Other undershirt-sweat research has shown that women prefer the scent of men whose genetic antigen/antibody system is different enough from theirs that if they were to mate with them, their babies would have the right combination of genes for a good immune system.

It's possible that a guy could cheat using this knowledge. If he wanted to attract a woman who was not attracted to him, he could find a guy who is attractive to that woman and wear that guy's undershirt before that guy put it in the wash. It wouldn't be a good idea though if he wanted to have a baby with the woman. It's likely the child would not be as healthy

Just as is the case with adults, babies with the most symmetrical faces are considered the most beautiful, but babies have other physical traits which we instinctively find cute and appealing. Rounded cheeks, especially, large eyes, small mouths and chins and little button noses make us ooh and ah, and naturally encourage loving, protective feelings. Scientists have done experiments where babies' faces were modified in a series of drawings, and they found that the smaller the eyes, the bigger the noses, and the more angular the features, the less appealing the babies were. Babies with big noses, sharp jaws, narrow faces and small beady eyes do not stimulate loving feelings or inclinations to cuddle. We don't love babies and find them appealing just because they are small and helpless and innocent. We instinctively love them and feel they are appealing because of the way they look!

Our natural inclination to find babies cute and appealing spills over to engender similar feelings toward puppies and kittens and other baby mammals with the same large eyes, rounded heads and cheeks and small noses and chins and other baby-like features. That's the main reason why pandas are so much more appealing than rats. People who design dolls and stuffed animals like teddy bears know, of course, that these features naturally appeal to us. So do some successful artists who deliberately exaggerate baby-like features in their drawings of adults and children. Clients who buy their art are usually unaware that the subjects portrayed have eyes that are a little larger and noses that are a little smaller and cheeks that are a little rounder than are true of real people. They just know they find these drawn portraits appealing.

Cartoon characters designed to be appealing, like Mickey Mouse and Dora the Explorer and hundreds of others are usually drawn with very exaggerated baby-like features. Very large heads relative to body size, large eyes, rounded cheeks, small noses and chins are typical. It's the same with the bunnies and ducks and other characters who greet visitors at Disney World. Evil or sinister characters like wicked witches, though, are drawn with small eyes, angular features, large noses and sharp chins so they will not instinctively appeal to us.

Most any stimulus which is naturally attractive can be made even more attractive if it is exaggerated and made supernormal. Drawings of sexy women such as the pinup girls on calendars (found in most mechanics shops and barbershops in the 1940's) showed them with legs often 30-40% longer than women's legs really are! Alberto Vargas's pinups were especially famous. It's doubtful whether many men noticed this distortion, much less objected. Nor do men complain when graphic artists greatly exaggerate the size of women's breasts. Supernormal stimuli are supertittilating.

Pictures and erotic drawings of genitalia are much more interesting to men than to women so it's not surprising that drawings which exaggerate female genitalia are attractive to straight men and those which exaggerate penis size appeal much more to gay men than to women, as did the Tom of Finland drawings which showed monstrous genitals several times normal size. Women do respond, though, to somewhat exaggerated depictions of masculine faces and bodies on the covers of all those paperback romance novels.

Habitats and landscapes

Among the many instinctive attractions all animals naturally respond to are features of the habitats they have evolved to live in. Bullfrogs are attracted to the edges of quiet ponds and lakes but not to rushing rivers or forests. Squirrels are attracted to large trees but not to habitats with shrubs and small saplings. Lions like tall grass which provides cover while stalking, and pillbugs (rollie pollies) naturally seek dark, moist habitats under logs and rocks. Animals know where to live because of their instinctive attractions and aversions.

What about us humans? Are we naturally attracted to any specific habitat features? Well, yes, it seems we are according to several research studies of human landscape preferences. The landscape features we find aesthetically appealing appear to have been shaped by our evolution, and this has stimulated a lot of discussion on the biology of art.

Denis Dutton in his book *The Art Instinct* reports on several studies which suggest that people all over the world from very diverse cultures show strong agreement when it comes to landscape preferences. It doesn't matter whether the people responding to the surveys live in cities or rural areas or deserts or dense forests; people prefer savannah-like landscapes with low grasses, clumps of climbable branchy trees, the presence of water, a view of the horizon and evidence of resources such as animal and green plant life including flowering and fruiting plants and those where there is the prospect of refuge.

Just as other animals instinctively prefer those habitats ideal for their survival several scientists have proposed that preference for savannah-like habitats like where we probably evolved for so long, may be instinctive now in us. One might think that people would prefer comfortable, familiar man-made landscapes such as cities or buildings or parks but this is not the case. Most people prefer natural landscapes especially if they are savannah-like. It is curious why people who live in such diverse habitats as Iceland and Denmark and Kenya and the Ukraine and Turkey and the U.S. would all agree on the same preferences if there's nothing innate here.

The bottom line when it comes to common attractions and aversions most people agree upon, is that what tastes good and what smells good and what looks good to us is (or at least, was) usually good for our genetic survival and reproduction, and what tastes or smells or looks bad has evolved because it helps us avoid what is not good for our survival and reproduction. That's why sugar instinctively tastes sweet

INSTINCTIVE ATTRACTIONS AND AVERSIONS

but rotting foods make us gag, why women like tall men with broad shoulders and rugged faces, why most men like young women with large breasts and with waists smaller than hips, why everybody thinks symmetrical people are more attractive, and probably why most people in diverse cultures prefer savannah-like landscapes over others.

CHAPTER 6

Concealed Estrus, Sexy Bonobos, Porcupine Dildos And Gay Rams
(Evolution and the Purposes of Sex)

A RECENT GLOBAL study of frequency of sex in many countries showed that the average human has sex with a partner 103 times a year. Greeks have the most (138) followed by Croatians, Serbians and Bulgarians. Maybe there is something in the water in that part of the world. Americans and Russians were high on the list while the Japanese were way at the bottom (45 times per year). These averages tell only part of the story though; so many factors influence frequency of sex. Young newly-coupled people, for example, have much more sex (with partners) than singles, old married people, and curmudgeons.

One thing is clear; humans have far more sex than most animals. We have more sex than 99.99% of all species! Some animals, like rabbits and mice, with short gestation times, may have sex as often as 10 times a year, but most animals have sex for a brief period only once a year. If your sex tally is nowhere close to 103 and you feel deprived, just be glad you're not an elephant. Because their gestation period is so long, elephants engage in sex only for a short period every four years! Even then, as is the case with most mammals, many of the males don't have any sex at all because the females are picky and mate only with the best genetic studs.

Humans in their late teens and early twenties often have sex 10 to 12 times per month and even more often if they have a steady partner. That's about 10 times more sex than even rabbits have. People in their forties and fifties average two or three times per week. That's about 100 times as much sex as most mammals have. People in their 60's and 70's vary a lot but if they average only once a month that's about 50 times as much sex as elephants have! I wonder if you showed a picture of a 70 year-old woman and a picture of an elephant to people, and asked them who has the most sex, how many would guess that old ladies have, on average, 50 times more sex than elephants.

The reason most animals don't have sex very often is because females come into heat (estrus) and ovulate only a few times a year (mice and rabbits), or more commonly, once a year (most mammals), or once every four years (elephants), and females will mate only when they are ovulating. That's the only time the egg can be fertilized and the only time females are interested. What would be the point of having sex at other times? Isn't procreation the purpose of sex?

Well, it may be the main purpose in most animals but this is certainly not the case with humans! **Most human sex has nothing to do with making babies!** Most human sex naturally happens when there is no possibility of pregnancy!

Until menopause at around age 50, women produce an egg once a month, and the egg must be fertilized within one day or it dies. Sperm can live in a woman's body for about 4 days, thus if a woman has sex within 4 days before ovulation, pregnancy is possible. So in a month of 30 days there is only a five-day period when pregnancy can result. That means most of the time a man and woman have sex there's no way an egg can be fertilized. What's going on here? Why do humans have sex when reproduction is impossible?

The answer is clear to evolutionary biologists; the proximate reasons why frequent non-reproductive sex is so common in humans is because of concealed estrus in women, and because women are biologically inclined to want to have sex when they can't become pregnant. Unlike other female animals, estrus is concealed; women do not show the males when they are ovulating. Other animals show with visual signs like swollen genitals or chemical signals such as sex pheromones. Human females, though, give no clues so men don't know when sex can result in pregnancy. Women usually don't know either, and also unlike most other animals, women naturally want to have sex even on those days (about 5/6 of the time when no egg can be fertilized). Bambi's mother certainly didn't want to have sex when she couldn't get pregnant. Why did this system evolve in humans?

Concealed estrus and continuous sexual interest when pregnancy is impossible evolved in human females because it was very advantageous to those women; it kept their men around for more sex tomorrow. This promoted mate-bonding and thus encouraged men to protect their women and provide food for them and their offspring. Concealed estrus and the continuous sexual receptivity of women helped human males evolve to be good husbands and fathers unlike Bambi's dad. One might say that all this non-reproductive sex men and women have has promoted family values.

CHOOSY WOMEN AND CHEATING MEN

All this sex did not make men strictly monogamous but there were important evolutionary advantages for males too, of sticking around and bonding closely with their mates. For one thing they got a lot more sex; no need for costly time-consuming courtships new females would require. Also, by having frequent sex with the same woman, men could watchfully guard them to prevent cuckoldry and could therefore be much more certain that their children were really theirs, not some other male's. And, being more sure their kids were theirs, encouraged men to help with food and protection, greatly increasing the chances that their children would survive. Concealed estrus didn't prevent men from Coolidge-effect promiscuity but it did lead them to be better mates and helpful fathers which in the long run turned out to be a great way to pass on their genes. Human fathers are the only primate fathers which play an important role in care and protection of their young

There is one species which has more sex than we do; it is our very close relative, the bonobo, and it is the only other animal species where estrus is concealed instead of being advertized. Humans, chimpanzees and bonobos all are descended from common ape ancestors who lived about 8 million years ago. We are equally related to bonobos and chimps, but partly because of concealed estrus, we are much more like bonobos than we are like chimps in the frequency of sex. Bonobos are also the only other primate species which copulates face to face.

Bonobos have sex many times a day and their sex has even less to do with reproduction than ours. Most everyone has sex with everyone else. Males and females have frequent sex with one another but females also have frequent sex with one another by rubbing their genitals together. Males often rub their scrotum against other male's buttocks or rub their erect penises together. There's lots of oral sex and tongue kissing too. Adults often have sex with juveniles as well, although it seems mothers do not have sex with their sons, thus avoiding incest problems. Because we are so closely related to bonobos, there may be some clues here for understanding human bisexuality and homosexuality and pedophilia.

Why do bonobos have so much non-procreative sex? Having all this sex does not increase the number of their babies. Females can produce only one child every five years or so. It takes that long, given nursing and extended childcare, before a mother can raise another youngster.

And why is there so much homosexual and bisexual and even pedophilic sex which produces no babies? What's going on? Well, it's because almost all sex in bonobos has evolved to promote social bonding. **Social bonding is the main purpose of bonobo sex just as it is with humans except bonobos use sex to bond with all members of the group not just their mates.**

Bonobos also use sex to reduce aggressive conflict. If there is likely to be a dispute over a resource such as a food source they all have sex with one another first to reduce competition. And if there has been a fight they make up afterward by having sex with one another. And, when a female leaves her natal group at sexual maturity and joins a new group where all are strangers to her, the first thing she does is have sex with most of them. Better than a nice smile and a handshake.

Because of all this sex, bonobo social interactions are much less aggressive than those of any other primate species. Bonobos make love not war. Also, the frequent sex between females has led to female alliances so strong that bonobos are the only primate species where it's females not males who are socially dominant!

◄ CHOOSY WOMEN AND CHEATING MEN

When sex evolved hundreds of millions of years ago, probably first in bacteria, its only purpose was for procreation but, of course, because it is a pleasurable drive, it has become co-opted for other purposes, especially social bonding as happened in humans and bonobos. Many other animals too use sex, even homosexual sex, to encourage social bonds. The two male lions (often brothers) which lead a pride often strengthen their loyalty by having sex with one another. Male dolphins in a pod cooperate intensely to locate females and protect the young. They too have sex with one another and it is believed this strengthens their bonding to one another and reduces aggression.

Homosexual behavior has been observed in about 1500 species of animals and is well documented in more than 500. While it commonly promotes social bonding and reduces aggression it is not always clear, though, what other purposes it may have. Take seahorses, for example. Seahorses, long thought to be monogamous with lifetime mates, have turned out instead to be very promiscuous and often have bisexual and homosexual relations. In one big study at 90 aquariums it was discovered that 37% of the sexual encounters were with members of the same sex! Female-female sex was twice as common as male-male. Is bonding happening here? It could it be that seahorses are just indiscriminately horny, but it seems much more likely that there is some evolved reason for this. Even when they aren't having actual sex, flirting is very common with up to 25 potential flirt-partners a day of both sexes. For a more complete discussion of homosexuality in humans, it's proximate causes, and hypotheses as to why it evolved see the chapter on *"Why People are Gay"*

Procreation and social bonding are the two commonest

CONCEALED ESTRUS, SEXY BONOBOS, PORCUPINE: DILDOS AND GAY RAMS

purposes of sex, but there are many others. For example females of many species often trade sex to acquire food or other items of value. Some studies have indicated that chimpanzee females are more likely to mate with males who bring them meat, and it seems likely that ancestral human females were more likely to offer sex to men who shared meat from the hunt with them. Some female penguins, even if they have a mate, offer sex to strange males who bring them the pebbles they need to build their nests, and some birds invite sex with males even after their eggs have been laid thus tricking them in to care for chicks they didn't father. **There are lots of reasons why it is natural and normal and very adaptive to have sex when no reproduction is involved.**

And what about masturbation? Is it done just for fun, or does it serve some adaptive purpose? Most everyone does it. Even the sweet old spinster who won't even say the word 'sex'. Even the priest who believes it's a sin.

It is very common in other animals too. The most usual method is by rubbing the genitals against other body parts or on the ground or on inanimate objects or other individuals. Dogs sometimes hump people's legs, even grandma, or the preacher making a house call. Birds do it, lions do it, zebras do it and so do deer and chimps and they don't go blind or get acne. Some animals use their hands or paws. Male goats and kangaroos and monkeys sometimes suck their own penises (autofellatio) resulting in orgasm. Females of many species including porcupines and orangutans, have been observed to use sticks and stones as dildos.

To an evolutionary thinker it is reasonable to presume that since so many animals do it, masturbation must be adaptive; it must have been favored by natural selection because it helps survival in some way. But how? Well, there are data that provide good clues. Masturbation in males removes stored geriatric sperm which have lost much of their vigor, and replaces them with younger more motile sperm which are better able to fertilize eggs. In female mammals, masturbation decreases the risk of infections by increasing the acidity of the cervical mucus and by moving debris out of the lower reproductive tract. So it seems like it's a good, natural, healthy thing.

It can also be helpful in avoiding unwanted pregnancies The governments of the United Kingdom and several European nations, in fact, recently collaborated to produce a pamphlet encouraging teenagers to masturbate. The motive is to encourage masturbation instead of having partner sex in hopes of reducing teenage pregnancies and AIDS. Masturbation also has probably saved millions of marriages. Men who jack off instead of having sex outside their marriages do not threaten their relationships.

Masturbation was definitely *not* encouraged by the nuns and priests when I was a student at Queen of the Miraculous Medal school, and at Saint Mary's High. Nobody used that word, but masturbation was definitely a sin. It was not clear, though, to most of us whether it was a mortal sin or a venial sin.

If you told the priest euphemistically in the confessional that you had 'touched yourself' or if you admitted that you had 'impure thoughts' (almost as bad as doing it), you would be told to say ten 'Our Fathers' and ten 'Hail Marys.' But what if you died before you had a chance to go to confession? Would you go to Hell? Millions of people have suffered guilt and fear because of these teachings.

The Roman Catholic Church still views masturbation as

sinful, as do some Protestant churches, Mormonism, and the Shi'a sect of Islam. The most common reason why is because of the belief that the sole natural God-given purpose of sex is procreation. But where did this idea come from? Biblical scholars say it partly came from a misunderstanding of the story of Onan in the Old Testament of the Bible.

Onan's brother died and Jewish law required that Onan marry his dead brother's wife so his brother would have descendants. Onan knew, though, that any children produced would be his, not his brother's, and besides he wanted to inherit his brother's wealth. So he refused to marry his brother's widow, and after having sex with her, he withdrew, and instead of inseminating her, the bible says he spilled his seed upon the ground, whereupon God struck him dead. Jewish scholars, and in fact most biblical scholars regardless of their personal religious beliefs, believe it is clear that God struck Onan dead because he refused to obey the Jewish law requiring him to marry his brother's wife, and because of his selfishness. The Catholic church maintained, however, that God struck him dead because he spilled the seed upon the ground (he had non-procreative sex). Based on that interpretation which most biblical scholars say is wrong, the Catholic church has long held that any sex which is not procreative, (masturbation, homosexuality, sodomy, etc.), or the deliberate prevention of procreation (birth control), violates God's natural law and thus is sinful.

But if the sole natural purpose of human sex is procreation, why were we designed, unlike most animals, to have sex with one another, when most of the time reproduction is biologically impossible?

Maybe the pope and the mullahs and the leaders of many churches should hold a big international conference to talk with evolutionary biologists to consider why humans have so much sex when pregnancy is impossible and to discuss social bonding, family values, acquisition of resources, masturbation, and the many other **natural** purposes of sex.

CHAPTER 7

Tall Bishops and Genuflection Genes

ONE OF THE first studies to compare height and occupation found that bishops were, on average, a couple of inches taller than parish priests. It also showed that sales managers were typically taller than salesmen, and that university presidents were generally taller than presidents of small colleges.

The tallest men of all turned out to be reformers, superintendents, wardens, and governors. They were about five inches taller, on average, than the musicians and publishers at the bottom of the list.

It's not clear why musicians and publishers were statistically so short. Maybe just chance. But It wasn't chance which determined who ranked at the top of the list. Since that research study was published by Gowin in 1915, many others like it have come to much the same conclusion: Men in positions of public authority and leadership tend to be significantly taller than average.

Most U.S. congressmen and other politicians, for example, are relatively tall. There hasn't been a president under 5'9 (the average for U.S. males) since McKinley won in 1900. In fact, the taller of the two major presidential candidates has won well over 80% of the elections since then! Obama was the most recent example. A gambler who knew nothing about the candidates except their height could have made a lot of money by always betting on the taller one.

It's not only in the U. S. that the taller candidates for high public offices are usually selected. It's true all over the world. From tribal chiefs in tropical jungles, to metropolitan chiefs of police, the general rule about rulers, men at least, is that they are usually taller than most of the men they rule. One can always find Napoleonic exceptions, of course, but most elected political leaders are more like de Gaulle. They're tall. Why is this so?

First of all, it's because of people like you and me. *We* are the ones responsible for this heightist state of political affairs. After all, bishops and union leaders and presidents are not born, like royalty, to high office. We choose them, and when we do, we usually vote for the taller candidates. It's clear we prefer leaders we can *look up to*.

Not that we realize this when we cast our votes. You won't likely hear any of the cardinals who chose the bishop say that his evaluation included points for height. We believe we select those candidates who are most qualified, and those we most respect. But it's clear we subconsciously feel that the taller candidates are better qualified. We respect them more *because* they are taller. After all, we usually vote for the ones who are *head and shoulders above the rest*.

Many studies have shown that status and stature are closely related in our subconscious appraisals of others. From Trobriand Island tribesmen to the top movie actors, taller men tend to be more respected and admired. Psychologist Paul Wilson found, for example, that college students told that a class visitor was a professor, estimated he was two and one

TALL BISHOPS AND GENUFLECTION GENES

half inches taller than did students who were told he was a student. No surprise here. The research has repeatedly shown that taller persons tend to be evaluated more positively, women it seems as well as men, though few studies have specifically compared female status and stature.

Height, of course, is only one of many factors which influence how a man literally *measures up*. The clergy charged with choosing a bishop will not select a heretic, no matter how tall he is. Nor will union members vote for a towering milquetoast. But all else equal, the taller candidate has a significant advantage. Short aspirants to positions of political power perhaps should be forewarned of our bias and their disadvantage. It might be best to encourage them to become musicians or publishers, or pursue some other such career where success depends so much upon ability that height makes little difference.

It's not just more political power and general status that taller men enjoy. We give them many other perks as well. There is much evidence, for example, that the taller applicants for *most* jobs are more likely to be chosen, even when the positions do not involve leadership. It would seem advisable to stand tall, and wear elevator heels even if one is applying for cook in a Chinese restaurant.

Once they get the job, tall men are more likely to be promoted, and they also tend to be better paid. Leland Deck of the University of Pittsburg showed that male graduates of that college who were 6'2" or taller had salaries 12.4% higher than those who were below six feet. Large recent American and British studies indicate that each inch is worth about $800 per year in 2008 U.S. dollars. Men who were 6'2, averaged about $8000 more per year than men who were ten inches shorter at 5'4. That's about $280,000 more in a thirty-five year career just for being *highly qualified*.

◄ CHOOSY WOMEN AND CHEATING MEN

Adding romantic insult to general injury, from the perspective of short men, the research also shows that taller men, the world over, are usually perceived by women to be more sexually attractive, they are preferred as escorts, and they are more likely to be chosen as mates. And, in societies which permit polygyny they typically have more wives. When it comes to choosing dates and mates, even feminists who rail against sexism usually practice heightism.

Of course there are many short men who have no trouble attracting women. Lots of evolved preferences for certain male behavior traits influence female choice as David Buss of the University of Texas has shown in "The Evolution of Desire". And human females, like females of all other species, have evolved preferences for certain physical traits as well. So, handsome faces, broad shoulders and muscular legs all help at first, at least across a crowded room or beach. Still, a tall handsome man is more likely to get a second glance than a short handsome man, and a tall ugly man more often gets a second date than a short ugly man. From singles bars to church socials it's the short men who are most often shortchanged.

Eventually, perhaps, short men, and women too, will form some sort of short-lib union to try to prevent this pervasive discrimination they've suffered for so long. But, in the meantime, what can they do?

Dishonesty is often the best policy. One old trick which often helps if it's not too obvious, is to make oneself look taller with soapboxes and pulpits, thick heels, raised hair styles, hats etc. 1940's movie star Alan Ladd legendarily overcame his 5'4 shortcoming to convincingly play romantic swashbucklers by standing on boxes while scenes were filmed. During strolling scenes on moonlit nights, the leading ladies walked in trenches

TALL BISHOPS AND GENUFLECTION GENES

beside him. Someone's sure to suggest that all shorter political debaters should be allowed, as Dukakis was, to stand on hidden soapboxes, and perhaps some fair-minded person will propose that the shorter candidates for bishop could be secretly propped up on pillows before the interviews.

But boxes and pillows will not be nearly enough. There's not much most candidates, job applicants, and guys in singles bars can do to appear taller without seeming pathetic. Everyone notices if a short man has four-inch heels. Ladd's deception succeeded only because his studio was able to hide the truth. If the real Alan Ladd had stood up in the first scenes, next to the leading ladies and other actors, most of them much taller, it's doubtful that boxes in the subsequent scenes would have helped at all, except to turn his movies into black comedies. Dukakis lost the first round in 1988 when he walked across the stage to shake hands with Bush.

If boxes and pillows won't do the trick nowadays, what else? Equal opportunity laws? Guidelines requiring cardinals to choose a representative percentage of short bishops? Affirmative action programs requiring that women date, and mate with, an appropriate percentage of short men? Get outta here! Most people don't even realize they seriously discriminate against short people, and if they did, few would admit it. **Our heightist bias is deeply ingrained. It's ancient and irrational and largely subconscious.** Laws won't help much as long as people pull the curtain when they vote, and as long as women can pick men they are instinctively attracted to.

Why do we respect height and give tall men such advantage? And *why* are most women more attracted to taller men? *Why* do we have this heightist bias? Is it something we learn? Some perversity of acculturation?

Evolutionary psychologists do not think so. There is no good evidence that our heightist inclinations have much to do with what we learn. The logic and evidence for biological causation though is very strong. It is mostly because of ancient *genetic* inclinations that we respect height, and women are attracted to tall men.

For one thing, we humans, the world over, all have much the same biases. I don't think there's even one exception—not even one culture from Pygmies to Norwegians where tall men aren't respected more and chosen more often as leaders and mates. It's a powerful argument for biological influence when most people in hundreds of diverse cultures all show the same inclinations. And, it's made even stronger by the fact that most of our mammal relatives also show respect for, and defer to taller individuals and give them advantages, and by the fact that most female mammals show preference for the biggest suitors as well. It's not acculturation which explains why mice and monkeys respect bigger individuals. And it's not learning which causes female deer and elephant seals to be more attracted to the larger males.

It may not be reasonable nowadays to pick bishops and boyfriends even partially on the basis of their height, but it's not surprising to evolutionary psychologists. We humans still have all sorts of irrational, and even maladaptive vestigial inclinations which still influence our behaviors today. Like our dogs who walk in circles before lying down, we are stuck with many leftover predispositions and programs which don't make much sense in suburbia.

To understand our human feelings about height then, we need to understand why natural selection commonly favors genes for respecting taller individuals, and genes in females for swooning over big males.

Evolutionary Explanations

In species where individuals physically compete for resources such as food or territory, it is usually an advantage to *be* bigger than one's competitors. In fact selection has favored genes for many mechanisms which make animals merely *look* bigger when involved in disputes. Raised hairs and feathers and puffed up bodies are common in situations involving conflict. By being (or looking) bigger, disputes are often won "paws down". Smaller individuals tend to deferentially step aside and let the larger ones have their advantages. "Yes sir, anything you say."

But why is this so? Why respect toward bigger individuals? Why don't the smaller animals fight instead for equal rights? Basically it's because respect is healthy. Unless one is backed into a corner, or options are otherwise limited, there is little to be gained by challenges which ignore the practical advantage larger individuals have. So selection has favored genes for respectful feelings toward bigger individuals, and for backing down and giving up, for the moment at least. This is true of lizards and lions and mice and monkeys and thousands of other animals. Since we evolved by natural selection, like our animal relatives, it's no wonder we too, have biological feelings of respect for tall individuals.

To explain why human females prefer tall men, scientists point out again that we humans are not unusual. Natural selection has produced similar biases in many other species. Females of most all mammal species, for example, usually choose the biggest males. This is especially true of polygynous animals (most mammals), where males physically compete for females. From deer to gorillas, a few males, usually the largest ones, mate with many females and many of the males never mate at all. This is not so much because the larger males prevent the smaller ones from opportunity, as because females of these species, like human females *choose* the larger males.

And here's why they do. Females genetically inclined to prefer the largest males, therefore have sons which tend to be larger, more effective competitors, more likely to survive and thus pass on the genes for their mother's preference to their daughters. So genes for swooning over the big bruisers are typical in females of polygynous species. Since our ancestors were quite polygynous for millions of years back, (a great majority of primitive human societies still are), it's no wonder that most human female hearts, like those of most mammal females, flutter more over tall males. It is not rational nowadays for a feminist concerned with equal rights to discriminate against short suitors, but we are, all of us, feminists too, irrational much of the time because of our primitive genes.

In the most polygynous species of all, such as elephant seals, where one male is harem master and does virtually all the mating, competition for females is so extreme that there is especially strong selection pressure for genes for large male size, and therefore female preference for the same. Thus in these species, males have evolved to be enormously larger than the females, and the females, predictably, show little interest in mating with any male other than the harem master who is typically the largest of all. If elephant seals had singles bars the biggest guy would go home with nearly all the females in the place.

Among monogamous species, foxes and beavers, and most birds for example, the situation is quite different. Some philandering aside, each male pairs with only one female, for the season at least, so there are more females to go around and less advantage for genes for large male size. So selection does not much favor female preference for especially large males. This is why males and females of most monogamous species have evolved to be typically equal in size. It's easy to tell when you are at the zoo which species are monogamous and which are polygynous. If the male and female are about the same size they are monogamous but if the male is bigger they are polygynous. Since human males are about ten percent bigger, on average, than females this suggests our ancestors were not monogamous like foxes and beavers but were polygynous like most mammals though not extremely so like elephant seals.

Genuflection Genes

What does all of this have to do with "genuflection genes?" Scientists who study the evolution of behaviors say that it has not been adaptive enough to merely respect larger males and to acquiesce without much contest, allowing them the top leadership positions, the prime cuts, the best territories, and the smitten females. It has not been enough to merely *be* intimidated and impressed. Natural selection has also favored genes for behaviors which clearly *show* respect. Communication of respectful deference saves animal's skins by making it clear no threat or challenge is intended toward the larger individual who might otherwise attack, and is more likely to win if he does.

So, many appeasement behaviors have evolved. Some animals, like wolves, lie vulnerably on their backs or bare their necks to dominant adversaries, and many, like rats, adopt a female copulatory posture to show they are subordinate.

These behaviors say "I give up—you win—I don't challenge you—I'm no threat." So then there's no advantage for the other animal to attack. It's not adaptive to risk one's neck to fight a rival who has already surrendered.

The commonest way, though, that animals show healthy respect is by crouching and flattening their ears and tails, or otherwise making themselves appear *smaller*. Since this is the opposite of large size, and bristling hairs, and physical inflations which naturally intimidate, it symbolically communicates subordination and lack of threat.

When a subordinate chimp greets a dominant individual, the dominant one stands erect and holds his head high and raises his hairs, while the subordinate chimp, with his head and body lowered, and with his hairs flattened, scrapes and bows repeatedly, and thus appears much smaller than he really is. It's quite like the situation where a king, raised on a throne, and made taller yet by his crown, is greeted by his humble subject, who bows or kneels, lowering his head and body before his *highness*.

This genetic inclination for body-lowering appeasement is very common among the vertebrates especially in the social mammals such as wolves and chimps. Being mammals ourselves, bishops and publishers alike, and since we share nearly 99% of the same genes with chimps, it's no wonder that our kings and bishops and Indian chiefs stand tall and exaggerate their height with thrones, and pulpits and crowns and miters, and headdresses, and so on, while we naturally communicate instinctive respect by lowering our heads and bodies in their presence, bowing and genuflecting and such.

We don't have specific genuflection genes, of course, nor do bishops have genes for wearing miters or kings have genes for sitting on thrones. Our cultures have variously shaped these

TALL BISHOPS AND GENUFLECTION GENES

surface forms of our natural inclinations to express intimidation and appeasement. But there are strong biological currents below. As Lumsden and Wilson indicated in *Promethian Fire*, our cultural traits are commonly on a genetic leash.

No one has to teach us to stand tall when we feel confident, or remind us to raise our arms in the joyous rush of victory. We have to learn a proper curtsy, sure, and to genuflect with one knee on the floor, but no one has to teach us the deep biological rules. We innately make ourselves smaller when we feel intimidated or respectful just as other animals do.

"Hang down your head, Tom Dooley", says the folk song, but Tom wouldn't need any advice about his posture if he felt sorry. When we feel ashamed or intimidated, when it's the best subconscious strategy to appease kings and gods and victors, etc., we instinctively hang down our heads like thousands of other species. Like a dog who shrinks out of the room with his tail between his legs when caught snatching the liver pate from the coffee table, people caught red handed and charged with undeniable blame, almost invariably appease by making themselves smaller before their judges and juries. You can tell a lot about who's dominant and who's subordinate, who's threatening and who's appeasing by a person's posture.

So while we don't have genuflection genes per se, genuflection, like the fact of tall bishops, and female attraction to tall men, cannot possibly be understood *without* reference to genes, and the evolutionary circumstances which favored their selection.

Our innate respect for tall individuals influences even our designs of buildings and monuments which house or symbolize important people and gods. Churches and castles, for example, vault with steeples and spires above their surroundings to impress us with the status of their occupants. A practical church or castle one or two stories high would be so much cheaper to heat and paint, and it would be much easier to change the lightbulbs, and repair the roof but we wouldn't be nearly so awed. It's no wonder that penthouses for the powerful typically tower above the lowlife of the city, and the offices of corporation presidents are on the top floors, and it's the upper crust who usually live in the mansions highest in the hills,.

I don't suppose that there are many C.E.O.'s though, who ever think about why they are on the top floors, much less imagine it has something to do with our biological nature. Nor do Shriners speculate about genetic influence when they don their tall fezes for parades and formal functions. Bishops don't wonder, either, why they wear their miters, nor when they get together at conferences, do they discuss the hypothesis that bowing has evolutionary roots. Devoted Muslims, when they prostrate themselves before Allah several times a day, are not aware that this cultural appeasement behavior is on a genetic leash. Victorious boxers and politicians don't think about why they lift their arms in victory, and people with low self esteem rarely think about why they slouch. We scarcely realize that we are animals, much less that our behaviors are greatly influenced by genes favored millions of years ago by natural selection.

Our Language Reflects Our Instincts

I have used several words and phrases such as *"look up to,"* and *"head and shoulders above the rest,"* etc., which betray our natural feelings about status and stature, but there are many more. Our languages all over the world reflect the

TALL BISHOPS AND GENUFLECTION GENES

same natural biases and inclinations. When we are exuberantly happy, confident, feeling successful, victorious etc., we say we feel *high*. *We are on top of the world*. Our spirits are *elevated*.. Even if we're not making ourselves physically bigger by jumping up, and raising our arms like prizefighters and football champs, we naturally feel bigger when we are winners just as our words say.

When we feel unhappy though, especially if we feel inadequate, rejected, or unsuccessful, when our self esteem is poor, we say we feel *low*. We say we are *down in the dumps*. So do people all over the world express these feelings with similar words which say they feel smaller. We say we feel *depressed*. To be depressed, of course, is to be made smaller.

Similarly, we say we respect people with *high standards* and *top ability*, and those who rise above their limitations. We like *uplifting* stories about individuals with lofty ideals who *pull themselves up from the lowest ranks to become upperclass giants*. We not only *look up to people* we admire, we often *put them on a pedestal*.

But we don't admire *small-minded* and *short-sighted* people very much, or those with low moral integrity. If we are cheated we say "that *little bastard* short-changed me" even if the culprit was not small at all. Or, we call him *a low-down cheat*. When we insult others we *cut them down*, and when someone behaves in an especially despicable manner we say "*How could you stoop so low?*" If chimps could speak, I think their words would betray the same deep feelings about height.

And if it had been chimps, instead of us to preempt the niche we dominate, if evolutionary pressures had made them a bit smarter, favored them instead of us to make the quantum leap to technological societies, they'd probably refer to us as '*lower animals,*' and their leaders, like ours, would likely be elevated on something like our thrones and pulpits and shoulders. They would probably exaggerate their size with intimidating headdress height displays like those our kings and Indian chiefs and bishops wear. And, their leaders would almost surely occupy the top floors of their buildings as well. Chimps already know innately that respect and intimidation are enhanced by height advantage. They raise their hairs when they try to dominate, and before making threat displays they often climb a tree or stand on something which makes them taller and more impressive to the others.

Just as we might predict from knowledge of our evolutionary heritage, God is elevated most of all in our imagery. He is almost always pictured above us in the sky. Some people imagine him at the top of a staircase flanked by angels who descend in order of their status. The devil though is imagined to be somewhere far below in the depths of depravity and is never portrayed as physically bigger or higher than God.

It's no wonder that the sciences of ethology, sociobiology, behavior genetics, behavioral ecology, and evolutionary psychology are growing so rapidly. These biobehavioral sciences are profoundly revolutionizing our self-understanding. After thousands of years of primitive speculation about the nature and meaning of our existence, looking (naturally) upward to the heavens for answers, thanks to revolutionary refinements of Darwin's theory of natural selection, and application of our improved understanding to the behavioral sciences, we are, at last, able to scientifically answer thousands of ancient 'why' questions about ourselves.

Even why shorter presidential candidates rarely win, why kings wear crowns, why women favor tall suitors, why we use words like 'depression' and why bishops are taller than parish priests.

CHAPTER **8**

Why Charity Begins at Home
(Kin Selection)

MOST ANIMALS NEVER get to meet their parents. Hatched downstream or crawling out of a cocoon, they're on their own from the start. No mom or dad anywhere. Tadpoles get no food from their mothers, baby oysters get no cuddling, and baby grasshoppers have no dad around to protect them or teach them how to jump. No parental care at all for most.

Where parental care has evolved, as in the birds and mammals, the parent is helping the survival of his or her genes (usually hers) in the bodies of offspring. One could put this more romantically, but parental care is basically a way genes help insure the survival of copies of themselves.

Parents who do care for their young have to have some way to "know" who their offspring are, and this is often by an automatic type of learning called imprinting which happens soon after birth. Wildebeest live in big herds, and because the calves are all born around the same time, there could potentially be confusion as to which calf is whose. Mistakes are very rare, though, because right after birth, each mother imprints on the unique smell of her calf. If they get separated crossing a river or while chased by lions, they are usually reunited because the mother recognizes her calf thanks to the imprinting. In any case, the mother will only give milk to her own calf. Even if her calf dies, she will not give it to an unrelated orphan calf whose mother has died.

It's much the same with penguins and other colonial birds. All the eggs are laid and the chicks hatch about the same time and the parents sit only on their own eggs and feed only their own chicks. Abandoned eggs and orphaned chicks have almost no chance of survival.

Why is this so? Why do animals care only for their own young? Why don't they instead behave altruistically for the good of the species? Well, the answer is pretty obvious to evolutionary scientists: natural selection can only favor behaviors which help reproduce one's own genes whether they are in one's own body or whether they are in the bodies of offspring. Since parents and offspring share 50% of the same genes (in addition to the genes we all have like the gene for two eyes and the genes for a heart), natural selection has favored caring for one's own young, but not the young of others who are not related.

But animals are often altruistic, as well, toward relatives who are not one's children. How to explain this? This altruism toward relatives who are not descendants bothered Darwin for years, so much so that he considered it a challenge to his theory of natural selection. He couldn't understand how selection could possibly favor behaviors which helps another individual at some cost to oneself. Especially disturbing was his realization that honeybees die in defense of the colony. How could selection favor altruism so extreme that death results? Darwin was also very puzzled about how sterile

WHY CHARITY BEGINS AT HOME

castes could evolve in the social insects such as bees, ants, and wasps. The queen does all the reproducing. All the rest of the females (workers) are sterile, and never have any offspring of their own. How could natural selection favor *not* reproducing?

Eventually Darwin resolved this conflict somewhat by proposing that natural selection might not operate only on individuals, but on groups of related individuals. Altruistic behavior, even kamikaze defense and sterile castes, might evolve by helping the survival of close relatives. Darwin was almost right about this, but he could not explain precisely how this might happen. That would have been impossible then because he didn't know about genes. The scientific world didn't know anything about genes until Mendel's earlier research proving the existence of genes was discovered in 1900. Darwin knew there had to be something like genes, but there was no proof in his lifetime, so he could only speculate about mechanisms of inheritance.

Finally, in a famous scientific paper in 1964, William D. (Bill) Hamilton figured it all out: all the questions about altruism to relatives, sterile castes, and kamikaze bees that worried Darwin. Hamilton's insights revolutionized our understanding of natural selection and made selection theory far more powerful and explanatory than even Darwin could ever imagine in his wildest dreams. If I could bring a couple of scientists back from the dead for one day it would be Darwin and Hamilton so they could talk together. After a conversation with Hamilton, I think Darwin would feel like he'd died and gone to heaven. The intellectual excitement would be so great Darwin might just die again before his day was up.

Bill Hamilton would probably start out by telling Darwin all about genes and how they influence traits. But Hamilton would tell Darwin much more than just about the nature of genes. He would surely tell him that **evolutionary biologists today regard genes (not individuals as Darwin thought) as the essential units upon which natural selection operates!** Ever since George C. William's influential *Adaptation and Natural Selection* and Richard Dawkins's books popularizing the idea, evolutionary scientists have realized that natural selection is a process which operates on the differential survival of **replicators**. Genes are the only biological replicators. Individuals (except for some clones) are not replicators. It is genes which are the units being selected for or against, not individuals as Darwin thought. **Individuals are survival machines for the replication of genes**.

As Hamilton explained this modern gene-focused way of thinking about how evolution happens, Darwin would surely be gratified to realize that because of this focus on genes, not individuals, as the units of natural selection, altruism to kin other than offspring, kamikaze defense, and sterile castes pose no contradiction to his theory of natural selection as he feared.

Hamilton showed that all else equal, the degree of relationship will be proportional to the degree of altruism which evolves. Parents and siblings who share 50% of the same genes should be expected to be approximately twice as altruistic to one another as they are to uncles and aunts or nephews and nieces who share only 25% of the same genes. Sister worker bees who share 75% of the same genes with one another (because of the peculiar genetics of social insects) are even more altruistic to one another. But they are not very altruistic to their brothers, the drones, with whom they share only 25% of the same genes. No wonder workers even die in defense of the colony, since the loss of one set of genes results in the survival of many sets of the same genes in their sisters. Sterility of workers has evolved because instead of reproducing directly with drones, which would pass on only 50% of their genes, by manipulating their mother the queen instead to make more sister workers, 75% of their genes are passed on each time.

The concept that altruism will evolve to the degree that individuals share genes is commonly referred to as kin selection and I have used that term because it is more familiar and intuitively understandable, but Hamilton actually called it Inclusive Fitness. Natural selection favors genes which do anything which promotes making more copies of themselves including those copies found in other bodies. In fact, the recipients of altruism don't have to be close kin by common descent. They only have to have the same shared genes. In theory, at least, we might predict that people with big noses or type A blood would be more altruistic to non-relatives who share these traits. Philippe Rushton of the University of Western Ontario has some very interesting data on this. I remember when he presented these data at a conference in a paper titled *Beyond Kin Selection*, Bill Hamilton was present and his jaw dropped a foot.

Kin selection theory, which focuses on the gene as the unit of natural selection, has finally made sense of many puzzling facts about animal and human traits which, before Hamilton, could not be answered scientifically. The evolution of menopause in women, but not men, is one which was described in the second chapter. Here are a few more:

- **Alarm calls** Many animals such as Belding's ground squirrels make warning calls when predators approach even though this often increases the risk to themselves. As the theory predicts, they only do this when relatives are nearby. A female ground squirrel, for example, will make a warning call when her offspring or her sisters (genetic relatives) are in danger, but she will *not* make the call to save her mate or her friend Bessie who are not related to her! When males become sexually mature they migrate from their natal group and join another colony where none of the others are related to them. As kin selection theory predicts, they do not make any alarm calls to warn these non-relatives.

How do these animals 'know' who their relatives are? As with the wildebeest it's due to instinctive imprinting during a sensitive period early in life. Ground squirrels imprint using scent on their mother and siblings during a sensitive period which lasts a week or two before they first emerge from the burrow. Because of this imprinting they behave altruistically to these relatives when they are adults and they will not mate (avoiding incest) with them when they become sexually mature.

If a pup, though, is removed from the burrow before the sensitive period and not returned until it has passed, he will not have imprinted on them and he will not be altruistic to his mother and sisters and will try to mate with them. If a non-relative female pup is introduced to the burrow during the sensitive period and removed right afterwards, she will be altruistic to them when she is an adult just as if they were her siblings, and she will not mate with the males she imprinted on even though they are actually not her brothers. Incest avoidance mechanisms like this imprinting have evolved in many animals because when close relatives mate, double copies of bad recessive genes often come together, reducing the survival of offspring.

Something similar appears to happen with humans. Several studies have shown that incest-avoidance is universal to all human cultures, is instinctive and, just as with ground squirrels, imprinting is involved. In fact children raised together the first few years of life usually avoid mating with one another even when they are not relatives. One study by Weisfeld and Czilli et al (2003) suggested that in humans as with ground squirrels, smell (olfaction) is involved in the incest-avoidance imprinting.

- **Helpers at the nest** Some birds such as Florida Scrub Jays, when good breeding territories are scarce, and the odds of getting one for oneself are poor, will stay home, and instead of trying to reproduce directly, will help their parents raise more brothers and sisters by incubating eggs and feeding their young siblings. Since brothers and sisters are just as closely related to oneself as one's children (siblings and offspring each share 50% of one's genes) it works out better under these circumstances to go with a fairly sure thing and pass on your genes by helping to make siblings, rather than to fight the odds by trying to make offspring.

- **Lion prides** A coalition of two to six males defends the territory against other males and the most dominant ones mate with all the females in the pride. If the males were unrelated, there would be extreme aggressive competition for access to the females, but they don't usually compete very strongly because they are typically brothers or half-brothers. Even though most of the matings will be secured by only one or two of the most dominant males, there is always the chance that subordinate males will get opportunities later, and in any case, even if they don't, some of their genes are being passed on by proxy by their brothers and half-brothers.

Kin selection also makes sense of why when males first take over a pride, they kill all the cubs born to the previous males. This makes evolutionary sense since they are not related to the cubs, and by killing them, the lionesses quickly return to estrus and can soon produce cubs fathered by the new males.

It's not just lions. Many animal males kill the offspring of other males of their species for the same evolutionary reasons.

Another fact of lion life which is explained by gene-focused modern theory is that the tenure of males in a pride lasts only two or three years before they are driven away by a new coalition of males. This appears to be because the female cubs they have fathered have sexually matured by then, and mating with them, their close relatives, does not produce good reproductive success because of incest depression of fitness. After two or three years it's in the male's best genetic interest to "allow' themselves to be driven away by a new coalition. They will have better genetic success by trying to take over a new pride where none of the females are related to them.

- **Mothers versus fathers** We take it for granted that mothers of thousands of species love and care for their young. Of course they do. But *why* is this so, and why do human mothers love and care for their children?

Clearly for the same biological reasons; similar brain circuits and hormones are involved. The feelings of love and the inclination to care for young are instinctive in all mother mammals and in mothers of many other animals. Maternal care passes on the mother's genes.

Because females are 'sure' their offspring are theirs, and because females have very limited reproductive opportunities, females have evolved to provide extensive parental care. Because males of most species have far more opportunities to reproduce and have no idea which young are theirs (even human males today, without DNA tests, can't be sure babies born to their mate are theirs), males of most species (birds are an exception) have not evolved to be as biologically inclined to provide as much parental care. It's not surprising then that in every culture in the world, women spend far more time and effort caring for children than men do.

- **Children versus stepchildren** There are many people who are just as loving to stepchildren as they are to their biological children, but, just as kin selection theory predicts, many studies have shown that parents are usually more altruistic to their biological children than to stepchildren. They give their biological children more financial support, as for a college education, they leave them larger inheritances, and they are far less likely to physically abuse their biological children than their stepchildren. Daly and Wilson at McMaster U. in Ontario have, in fact, shown that a child living with one or both stepparents is 40 to100 times more likely to be fatally abused than a child of the same age living with biological parents! It's usually the stepfather, the male, not the stepmother who kills chidren who are not biologically theirs. This is quite like the situation with lions. The evolutionary shoe fits here doesn't it? Common folk wisdom is perhaps as interesting here as the scientific data. Versions of the Cinderella story are found in many cultures

- **Grief** Emotions like anger and fear have evolved because they are highly adaptive. It is not quite so clear, though, why we experience grief at the loss of a loved one. Humans, of course, experience grief very intensely but chimps and elephants also seem to grieve. It is such a powerful emotion that one might suppose it must have been favored by natural selection because it promoted genetic fitness, but there is a lot of debate about exactly why. Whatever the reason, the circumstances under which we feel the most and the least grief are rather neatly predicted by kin selection theory. For example when we equally love a close friend and a close relative, such as a niece or nephew, we usually feel more grief over the loss of the relative, especially if the deceased was young enough to have reproductive potential. Our genes in their body are lost.

    As adults, when our parent dies, the grief is not nearly as great as when our child dies. The death of a ten-year-old child causes far greater grief than the death of a parent we have loved for 50 years. The parent is not going to reproduce any more so no genetic fitness is lost, but the death of a child means real absolute loss to our future genetic potential. Just as kin selection theory predicts, the only other situation where people report grief so intensely is with the death of an identical twin.

- **Grandparents** If one didn't know about kin selection theory, one might expect that, all else equal, relatives on the mother's side and those on the father's side would be equally altruistic to the couple's children. But all else is not equal. It turns out in diverse cultures all over the world it is the mother's relatives, her parents and brothers and sisters, who are usually the most altruistic to the children. They tend to give more love and care and resources to the children than the grandparents and uncles and aunts on the father's side. This is because relatives on the mother's side can be 'sure' her children are their genetic relatives while relatives on the father's side cannot be certain they are related to the children. They could have been fathered by some other man. Not that anyone is consciously thinking about this; it's instinctive to feel more altruism toward individuals who are definite relatives rather than probable or possible relatives. Natural selection has produced these inclinations over millions of years.

    Kin selection theory has revolutionized our understanding of so many questions about animal and human behavior. Too bad Darwin and Hamilton will never have the chance to talk with one another about it.

CHAPTER 9

Vampire Bats, Mother Teresa, and True Love
(Reciprocal Altruism)

IN THE PREVIOUS chapter we saw that natural selection typically favors altruistic behaviors toward genetic relatives, and these behaviors could as well be called selfish as altruistic because they promote the survival of copies of one's own genes. What about those many behaviors, though, where animals and humans are altruistic to non-relatives?

Many animals, from vampire bats to chimps, share food with non-relatives, as do successful hunters in primitive societies, and many animals groom non-relatives to remove parasites. Baboons and chimps and humans form friendships with non-relatives and help them in many ways as well. When two male baboons, for example, are fighting with one another to gain access to a female in heat, a third male will often risk injury and take sides to help one of them mate with her though he himself does not mate with the female.

Are these examples of pure unselfish altruism? Usually

not, according to sociobiologists and evolutionary psychologists. The individual who is helping a non-relative is actually helping himself. If you watch for a while you will see that the helper will later be repaid when he is the one who needs help. It's called reciprocal altruism (RA). You scratch my back and I scratch yours. The idea was first formulated by Robert Trivers of Rutgers University, and many research studies and observations have shown that reciprocal altruism is very common in social animals and human interactions.

Vampire bats feed exclusively on blood and if they don't receive a blood meal at least every three days, they are likely to die of starvation. So they depend on RA to survive. If one bat gets lucky and finds a blood meal he shares it (by regurgitating) with a neighbor who was not so lucky, and when he, himself, is out of luck and can't find a victim, his neighbor will pay him back by returning the favor. Of course this reciprocation only works with bats who associate often with one another and remember who helped them before and who did not. Otherwise there would be too much opportunity to cheat; to accept help but not repay.

For reciprocal altruism to evolve it is not necessary that the goods or services exchanged are identical like blood for blood, as is the case with vampire bats. All sorts of goods and services can be exchanged. For RA to work it is only necessary that the benefits to each party exceed the costs. Female chimps trade sex for meat. Cleaner fish receive food (parasites) by cleaning their clients, and their clients benefit by having the parasites removed. Doctors perform surgery and patients pay them money.

The social life of chimpanzees provides many examples of RA. Reciprocity, for example, is critical to the maintenance of political coalitions, though the basis for female coalitions is different from that of males. Female coalitions are based on kinship and friendship and are usually stable for a long time, while male coalitions are based more on political advantage. Males form bonds with others who can help them to seek dominance and control (and ultimately, access to females). Male coalitions thus tend to be much briefer and opportunistic and depend more upon the shifting political winds.

For males to dominate it is not enough for them to be physically strong and intimidating. They must also be supported by the females as well. Without the respect of the females, a male cannot become a leader in the troop. And, when, with the help of the females, a male does become leader, he is expected to reciprocate by helping the social life of the troop to be relatively calm and orderly. He must act as peacemaker by intervening impartially in fights and taking the side of the underdog or he will no longer be supported by the females.

Males know they need female support to attain power, so when they anticipate a fight with a rival male, just before the encounter, they groom the females whose support they are going to need, and they often play with their children too. Politicians kissing babies!

When a chimp helps another in a fight, he expects reciprocation from that individual when he, himself, needs help in conflicts. If he is not given that help in his time of need, he often becomes moralistically aggressive toward that individual who did not repay, screaming at him or her, and chasing and hitting as punishment. Chimps also will beat up members of their own coalition who treasonously give aid or even kindly comfort to their enemies.

It is not only chimps who punish defectors who fail to repay altruistic acts. We do too. When we lend people money and they don't pay it back, when we babysit a neighbor's kids but she is always too busy when we need a sitter, or when we 'groom' a friend with compliments but he never says anything nice to us, we get moralistic and resentful too. "You owe me!" and we often write them off. "No more help for you buddy." We are very good at detecting cheaters who fail to reciprocate and we punish them by withholding future aid. Sometimes it can be over silly or petty issues where nothing much of value is involved. If I send you a Christmas card and you don't send me one next year, you may be cut off in the future. If I buy you an expensive birthday present but when it's my birthday you give me a fancy-wrapped vase you bought at a garage sale for $1.00, don't expect a nice gift from me next time. You're a cheater, Uncle George. Even three-year olds understand these things. In experiments they refuse to give toys to children who previously would not share toys with them.

Robert Trivers, while still a graduate student at Harvard, was the first to develop the theory of reciprocal altruism and he realized that many of our emotions and feelings such as gratitude and guilt probably evolved in humans partly because of our long history of reciprocal altruism. Also our sense of obligation, righteous indignation, and even our sense of justice. Guilt uncomfortably reminds us when we have not reciprocated or have been socially unfair, and thus keeps us from being too self-centered and ostracized by others, and the emotional display of guilt and shame shows others that we are aware of our social failings, and that we are contrite. More broadly, even our very sense of what constitutes moral and ethical behavior according to some evolutionary philosophers such as Wilson and Ruse (1985) has evolved because of our history of reciprocal altruism. **Moral and ethical behaviors evolved because they serve to reproduce our genes. Put this way moral behavior derives not from noble dispassionate high-minded concern for the good of all but is just another form of 'selfish' behavior. Robert Wright describes many of these ideas in his wonderful book, The Moral Animal.**

So much of human behavior involves reciprocal altruism. I invite you for dinner and you bring wine or flowers or dessert and/or invite me next time. I water your plants while you are on vacation and you bring me a gift. I drive our kids to the beach this time and you drive next time. I help you with computer problems and you repair my fence.

Are these examples of pure altruism? Are we helping one another just because we are kind and loving? Well, we may be kind and loving and offer our help without expecting any reciprocation, but if we expect to be paid back, or we stop helping when our good deeds are not repaid it's questionable how purely altruistic we are. Does a cleaner fish remove its host's parasites because it is such a kind and loving fish? Does the dental hygienist clean our teeth because she is so kind and loving?

I certainly do not want to suggest that all behaviors which appear to be altruistic are actually self-serving, much less suggest that they should be called selfish. Who would want to say that a mother who cares for her young is selfish because she is helping reproduce her own genes? Who would want to call a neighbor selfish because when he watches your dog he subconsciously, at least, expects you will gratefully reciprocate in some way? On the other hand, when a person is helping his own genes or expects to be repaid for his kindness should we call that person altruistic? How should we think about Mother Teresa?

Mother Teresa

Mother Teresa was very much admired and revered because of her extreme selfless help for the poor and homeless and blind and sick people of India who rarely received help from anyone else. Though poor herself, she devoted her life to helping others in desperate need, and she was awarded the Nobel Prize for her lifetime of selfless sacrifice. Was there ever anyone more altruistic than Mother Teresa? Can't we at least agree that *she* was very altruistic?

It depends on what motivated Mother Teresa, and we can never know that for sure. If she was unusually empathic and loving and had no expectations of reciprocation (which seems unlikely), then you could say, yes, she was truly altruistic. But if (as seems to be the case) she expected compensation for her good works, then it's questionable whether she was as truly altruistic as many assume. One *could* argue instead that she not only expected compensation but that she was extremely selfish, always thinking about what was in *her* best interest. Didn't she expect that she would live forever with God in Heaven? Forever! What's one lifetime of sacrifice compared with eternal happiness? What a deal! **One could argue that Mother Teresa was one shrewd lady; she calculated the costs and benefits and realized, selfishly, that reciprocal altruism with God was the deal of a lifetime. And she got the Nobel prize to boot!**

What about others who are motivated to be altruistic and do good deeds because they expect they will be rewarded by a place in Heaven? And what about atheists who perform the same selfless acts but do not believe there will be any Heaven for them in the hereafter? Should they be considered to be more altruistic than believers who expect to be rewarded?

It is usually much easier to recognize selfishness than altruism. When a person takes more than his share, or cuts into line, or repeatedly fails to reciprocate, it's pretty obvious and we feel righteous indignation or anger. When we are waiting in a long line of cars because of construction ahead and some drivers use the empty left lane or shoulder to try to get ahead of the rest of us who are patiently waiting, the outrage at this cheating is so great that some drivers will pull off to the left to try to block the lane and foil the cheaters.

The reason why altruism is more difficult to recognize is that we often don't know other people's motives. We often can't know for sure if their altruistic behavior springs from pure empathic unselfish goodness, or whether they expect reciprocation, or have some other motive which is self-serving. Compounding the difficulty of recognizing pure selfless altruism is the fact that people often do not even consciously recognize their own motives. We are all self-deceptive and frequently hide our selfish motives from ourselves, rationalizing and justifying our behavior to fool others and make ourselves feel better about ourselves. Self-deception is a natural part of human behavior and it has evolved because it is frequently necessary for us to lie, and the most effective liar is one who believes his own lies. Dishonesty is often the best evolutionary policy.

The bottom line, when it comes to behaviors we ordinarily think of as altruistic, is that they are often self-serving, either because we are helping our own genes, as with kin selection, or because we expect, and usually receive reciprocation, the benefits to us are at least as great as the costs

True Love

So is there any true love out there? Do some of us, at least some of the time, transcend our evolutionary past and behave in ways that are truly altruistic? Do we sometimes behave in selfless ways that do not help our genes, or do we behave in ways that are costly to us even when there is no hope or expectation that we will ever benefit? Of course we do! We sometimes risk our lives to save strangers from a burning car or house. Or we jump in a river to save a stranger from drowning. We volunteer at considerable cost to ourselves to help victims of disasters like Katrina or the earthquake in Haiti. We contribute to charities for homeless children in Africa. We donate blood to strangers. And we donate funds to green organizations to make the earth healthier even when we have no descendants and we are so old and sick we will not live long enough to enjoy a healthier planet. We volunteer at soup kitchens and give money to the humane society though the dogs and cats will never even thank us. We feel sorry for the victims of disasters and grief for the parents of children who die. We even cry in movies for characters we know do not really exist. We do these things because we, most of us, are naturally empathic. We humans are wonderfully capable of true love.

It is questionable how empathic dolphins or elephants or most other animals are. When dolphins try to rescue a swimmer or snorkeler who appears to be struggling it may be a mistake. Dolphins help struggling members of their pod by lifting them so they can breathe. It could be that when they help a swimmer, it is not because they are empathic and concerned about the swimmer. Instead it might be a misfiring of a natural evolved behavior to save any struggling member of their group, and they don't consciously realize the swimmer is not a dolphin. When elephants appear sad when a relative has died, or when they gently touch the bones of deceased members of their group it could be that they are experiencing empathy or grief, but it's difficult to know for sure.

It is more clear that chimpanzees experience genuine empathy for one another. They sensitively understand what others are feeling, and they clearly show sympathetic concern for one another. When a chimp is injured or sick, or has been beaten up or suffered a loss of status and is feeling bad, other chimps are quick to console by grooming him, or kissing him or putting their arms around him. "There there, don't feel so bad. I understand your pain and I care about you." Does this indicate genuine empathy? Is this true love? If so the origins of our exceptional human empathy and sympathy and love probably go back at least as far as our ape ancestry.

The distinguished primatologist Frans de Waal has written a wonderful book called *Good Natured* about empathy and sympathy and the origins of moral and ethical behavior in animals and humans. He argues convincingly that morality is grounded in biology, and that ethical behavior is a result of evolution. Another fine book which explores many of these questions is Nigel Barber's rich and fascinating *Kindness In a Cruel World*.

Our extraordinary capacity for empathy and love may have begun with the bonding of mothers and babies and it is nourished by cuddling infants. It turns out that touching, caressing, cuddling and other such body contact is critical for normal healthy physical development and for social development as well. Harry Harlow's classic monkey studies showed this decades ago, and much research has confirmed it since. Harlow's monkeys raised without mothers were much more hostile and aggressive and socially incompetent than monkeys raised with normal mothering. And, interestingly, when provided with dummy mothers they showed an instinctive need for body contact with *soft* dummy mothers greatly preferring the cloth mothers which supplied no milk to wire mothers with milk. It's no wonder little children when distressed, if the mother is not handy, often seek the comfort of teddy bears and soft plush animals and favorite blankets. Physical closeness in infancy promotes trust and emotional closeness, and is basic to later development of empathy and altruism. I used to tell my students if they forgot everything else they had learned in my Evolution and Behavior class that it is important to frequently hug babies if they are to develop well socially and emotionally. (Not that they needed to be told this—not the women anyway—parenting behavior is instinctive in us just like it is in all the other mammals).

Touch is extremely important in bonding in humans and in other animals as well. The licking of babies by mother mammals establishes the beginning of trust of the mother. And just as we bond with our babies with cuddling and caressing, we continue to bond with family and friends with hugs and kisses and handshakes for the rest of our lives. And, of course, we spend a lot of time petting our dogs and cats and they in turn sit on our laps and nuzzle our faces and rub against our legs. **It's touching how we bond with one another. It's significant too that we call our dogs and cats pets. To pet is to touch affectionately.**

Besides early cuddling there are many other variables which influence how empathic and loving a person turns out to be. Gender is one of them. Women are generally more empathic and socially sensitive (at least to the needs of family and friends) than men are. Blood levels of hormones and neurotransmitters such as oxytocin and serotonin are also important. Oxytocin, for example, promotes social bonding, and serotonin helps reduce aggressive impulses. Violent criminals usually have low serotonin levels.

Genetic inheritance can also make a big difference in how empathic and loving a person is. Inclinations to criminality are significantly heritable. and the extreme lack of empathy for others shown by sociopaths is largely due to their biology. Sociopaths, it seems are born not made. There is no evidence that how they are raised makes any difference, and there is no therapy which has ever been shown to be effective.

Some people are much more empathic and sympathetic and loving than others. It would be interesting to know how much of this difference is due to the influence of early parenting and how much is due to genes they inherit.

Thanks to the many studies of animal behavior and the revolution in our understanding of natural selection which began in the 1960's, we have come a long way in understanding ourselves. If we want to understand when and why we are selfish or altruistic, and why we are moral and ethical beings, or the roots of true love, best not to spend much time consulting the philosophers of centuries ago, or even the psychologists of a few decades ago. Better to consult the primatologists and sociobiologists and evolutionary psychologists of today.

CHAPTER **10**

What's Fun and What's Done

What's fun?

MOST OF US have seen chipmunks or other little rodents chasing one another on the lawn or forest floor. First one chases the other, then the roles are reversed and the chaser becomes the chased. It sure looks like they're having fun. This is not serious run-for-your-life chasing and running, as when a lion chases a gazelle. They are playing. But why play at all? Distracted by this apparently frivolous game they may fall off a cliff, or a fox or hawk may catch them. Why are they playing? What is play?

Play behavior is usually easy to recognize but not so easy to describe because there are so many different types of play. Here are a few:
- A kitten batting a wad of paper across the floor
- A little girl playing with her dolls
- Bear cubs rough-and-tumble alternating mock dominance and submission
- Scary thrill-play of people screaming on a roller coaster

- Men playing football or hockey or soccer
- Kids twirling around in the yard until so dizzy they fall down
- Playing cards and board games
- Play wrestling by gorillas
- Boys playing cops and robbers or violent video games
- Red deer playing "King of the Castle"

Most play is with members of the same species, but sometimes animals of different species play together, even predators and prey. A raven will sometimes peck at a wolf or dive-bomb him, then land in front of him inciting the wolf to lunge for the bird until the wolf is only a foot or two away, whereupon the raven escapes by flying off. "Ha Ha! You can't catch me" And then, adding insult to insult, the raven often returns to tease the wolf again! Similarly, a squirrel will often come most of the way down a tree and tease a dog nearby by chattering and moving its tail excitedly. Riled up, the dog barks and jumps to try to catch the squirrel, but the squirrel is careful to stay tauntingly just barely out of reach. When the dog finally gives up and wanders off, the squirrel will sometimes come all the way down the tree and even teasingly approach the dog on the ground until he is chased back up the tree, and the game resumes. Again, it looks like "Ha, Ha, You can't catch me". Why is the squirrel doing this instead of resting safely higher up, or eating acorns, or doing something else which doesn't risk its life? Dogs sometimes do catch squirrels and the game is over.

Is the squirrel having fun? Hard to say, but this behavior is surely pleasurable, at least. Does the squirrel "know" this is a game? Does the dog? Probably not. It is unlikely that either of them is consciously aware of why they are doing this. These behaviors are programmed. Squirrel mothers don't teach their young to tease dogs, and mother dogs don't teach their pups to chase squirrels; these behaviors are instinctive.

Animals know when others want to play There are many gestures and postures which basically say "Let's play." Dogs and wolves, for example, invite with a "play – bow" lowering the front of their body, extending their front legs, raising their rump, wagging their tail and barking. You might want to try this with your dog. Assume the play-bow posture and say "Woof Woof" to your dog. If you are very sober-sided and never play like this with your dog, he may nervously whine but more likely, he will recognize your invitation to play and will wag his tail. Chimps signal interest in play by tickling one another or by exaggerated silly movements or by showing a play face with an open mouth rather like a human grin.

What Do We Know About Play?

- It is pleasurable. Humans and other animals enjoy playing. As a rule, behaviors which are pleasurable, like sex or eating have been favored by natural selection because they are adaptive.
- Young animals play the most
- Play is correlated with intelligence and sociality: the most intelligent and social animals such as dolphins, chimps and bonobos and humans play the most.
- Play looks like frivolous behavior in that it doesn't seem to accomplish any immediate, necessary purpose.
- Play happens only in familiar safe environments. In new unfamiliar environments animals may cautiously explore but they do not play
- Play is costly in terms of energy expenditure and is

potentially dangerous. Animals often suffer injuries during play, and distracted by fun, may be killed by predators.
- The only animals which play are mammals and a few birds (You won't see grasshoppers chasing back and forth, or frogs taunting snakes)

So, what is play? Why do we play? Most scientists who study animal behavior believe that play is basically practice behavior. Animals fine-tune reflexes and motor programs in a safe environment, practicing behaviors which will later be critical for survival. The chipmunks are practicing running from predators, the squirrel is learning about dogs, the wrestling cubs are practicing dominance and submissive behaviors, the little girl is practicing being a mother, the kitten pouncing a string is practicing catching small animals and holding them down, the boys playing violent video games (which most boys love) are practicing competitive combat, and the men playing football and soccer are instinctively practicing group-against-group aggression (even though they don't realize it). Play is an instinctive program for learning.

The reason why it's mammals, and not most other animals who are programmed to play, is because mammals are much more dependent on learning. They need to learn more than other animals, and because they have the luxury of a long childhood protected by parents, this allows for the opportunity to practice behaviors which will be necessary for survival as adults.

What's Done

Think of a popular song which you, for a while at least,

WHAT'S FUN AND WHAT'S DONE

enjoyed so much you really grooved on it when you heard it on the radio, and were enthusiastic enough to download it or buy the cd. It doesn't matter what style it was, anything from country to rock. How many times did you hear the song before you liked it to the max? I have asked hundreds of students this question and they said they liked the song the first time they heard it, but they liked it more the second time than the first, and the third time more than the second, and most agreed that it took about five or six times listening before they enjoyed it to the maximum amount, and it was around this time, if they were to download it or buy it, that they did so.

The students also reported that they continued to enjoy the song more or less as much for another ten or twenty times but if they heard it many more times than that, their interest declined, and if they heard it much more, they finally became so tired of it they didn't want to hear it all. Please, "No More I Love You's" no more times.

What's going on? Why does it take a few times hearing a song to enjoy it maximally? And why does a song which was so enjoyable, even thrilling, if heard enough times, eventually become unpleasant to listen to any more?

Well, this illustrates how we are instinctively programmed to learn. When presented with a new unfamiliar phenomenon, a song or most anything else, we are instinctively wired to experience pleasure in learning about it, but once we have learned all about it we are instinctively programmed to lose interest. That frees us up to learn other new things. Otherwise we would be stuck grooving on the same song, so to speak, and we wouldn't learn much. Been there, done that. We instinctively, automatically become bored. Boredom has evolved to keep us learning.

It's no wonder why new gadgets, new hobbies and new

places are so much more interesting at first, and why we tend to tire of them after a while. We spend so much money on the pool table but then it sits in the basement, and is rarely used. We fall in love with Santa Fe or the Wisconsin Dells, or wherever, when there for the first time, and may start planning to visit again next year. But of course, next year is not nearly as interesting and if we visit several times more, we may still enjoy looking at the scenery and visiting the same shops and going back to the same restaurants, but the exploratory enthusiasm we felt the first year is pretty much gone. It's usually not a good idea in the first year you visit a new place, to buy a time-share which can't be traded for a new location. Better rent rather than buy that electric guitar the first few months just in case the usual natural boredom occurs.

Old Dogs and New Tricks

If you introduce a new object into a colony of chimps, nothing special, just something they haven't seen before—it could be a cardboard box or a roll of toilet paper— it's mostly the youngsters who show interest in playing with it and checking it out. The oldest chimps are much less interested, and often don't even bother to investigate it at all. They are not nearly as curious as the youngsters. Why is this so? Well, basically it's because curiosity killed the cat. Exploring new things and poking your nose into new places can be dangerous. There could be a spider in the toilet paper or a snake in the box. Why take chances?

So the old chimps aren't taking risks by exploring new things but the youngsters certainly are. Why the difference? It's because young mammals have no choice—they are bio-

WHAT'S FUN AND WHAT'S DONE

logically programmed to be curious. Youth is the time when most learning takes place. Sure, exploring new things can be hazardous. Curiosity sometimes has killed cats and often kills kittens, but youngsters are instinctively forced to be curious and thus take the risks. Otherwise they wouldn't learn.

It's very different, though, for old chimps and other mammals—they instinctively become neophobic. Neophobia literally means fear of new things but psychologists and animal behaviorists use this term also to describe what's happening with these old chimps; it is not fear of the new objects brought into the cage, but it is the reluctance or lack of interest or ability to explore and learn about new things and it has evolved to be typical of old animals. Once an animal has survived the hazardous exploratory curiosity of youth and has learned the ropes, it becomes increasingly counterproductive to be interested in what's new; it's better to play it safer and stick with what you know. That's why old dogs don't learn many new tricks. Nor do chimps and most other old mammals. We humans usually become neophobic too, even though we depend upon continuous learning more than any other species.

Check out the musical preferences of the old people in your community. They're not listening to Lady Gaga or Eminem or rap or hip-hop. They don't have any cds of Smashing Pumpkins or Culture Club or the Police, or Def Leppard, or Jefferson Airplane or the Doors, or even Tears For Fears. Most of them don't listen to any rock at all. In fact many of them don't have any music CDs though they have a record player in the basement.

Many of the people over seventy don't even like Elvis Presley or the Beatles much less Pink Floyd or The Moody Blues. They had already become too neophobic in their thirties and forties to appreciate these new sounds. Give them

Glenn Miller or Frank Sinatra or Perry Como or Connie Francis, maybe even Lawrence Welk. Most old people like the music they listened to when they were young and don't usually much like the music that the younger generations prefer.

The natural inclination to become neophobic with old age also explains why I-pads and I-phones and Blackberries are used mostly by young people, why 12-year-old grandchildren are usually more interested in and knowledgeable about computers than their grandparents are, and why, as Thomas Kuhn (*The Structure of Scientific Revolutions*) observed, new ideas, no matter how good or well-proven they are, become accepted only after the old generations die off and are replaced by younger people.

So, it is instinctive to play more when we are young, the type of play each species (humans included) engages in is largely instinctive, it is instinctive to become bored with what we have already learned, and it is instinctive to be less interested in learning new things as we get old. In fact most of what we call *learning*, from conditioned responses to language learning, is actually *instinctive*. Ask any ethologist.

CHAPTER 11

Homo Vestigius

WHAT WOULD HAPPEN if a Boeing engineer submitted a design for a jumbo jet with a little old-fashioned propeller just like that on the Wright brother's plane? And what if his plan included wooden wing struts and many other parts which were standard parts of the first airplanes, but which would be useless at best on a modern jet? What would the reaction be?

Unless he'd shown previous signs of derangement, his bosses would probably assume that he was joking. Perhaps they'd be amused, expecting some method to his madness soon to be revealed; some point he was humorously trying to make. In any case no one would take him seriously.

Nor should we take our own design as evidence of a serious plan according to many distinguished scientists and philosophers ever since Darwin. For one thing we have many badly designed structures such as our backbone. And for another, we are quite like jumbo jets with old – fashioned propellers and wooden wing struts. We have all sorts of useless body parts left over from our past evolution.

Biologists call them **vestiges**. More than one hundred have been identified so far. They were useful to our ancestors, often for millions of years, and are adaptive still for many of our animal relatives, but in us they are excess baggage at best.

We all have tails and gill slits and fish-type kidneys during our embryonic lives, for example. Even Jerry Falwell and Pat Robertson. And as adults we have many other vestiges such as our tail bones, and remnants of the muscles which once moved our ears to cup the sound and communicate. The same ones a dog uses to perk up his ears if he hears an unusual noise, or a cat uses to flatten them if you yell at him for shredding the couch. Unlike dogs and cats and most mammals though, our ancestor's ears became rigidly fixed thousands of generations ago and can't be moved now to catch sounds or communicate. It's true a few uncles can still use the remnants of our ear-moving muscles to amuse nephews and nieces with feeble ear wiggles. But that's about it. Like our embryonic fish kidneys and tails, they persist though useless now except as testaments against inspired design. They are fleshy fossils, evidence instead of our evolutionary past.

It's not just us humans who have vestiges. It seems all other animals and plants have them too. Vampire bats, for example, still have grinding molar teeth, though they now feed exclusively on blood. And while modern horses are one-toed and pigs are two-toed, remnants of the other toes still exist in them though it's been millions of years since their ancestors had five digits. Similarly, boa constrictors and whales still have imbedded leg bones, and dolphins during fetal life, have limbs which start to develop but then regress showing their ancestors lived on land. And, many species of blind cave animals have retained nonfunctional eyes. Of what use are a lens and retina and all the other parts of an eye to a cave fish whose whole life is spent in darkness, and which in any case, remains blind even if raised in the light?

HOMO VESTIGIUS

Among the most obvious of our human vestiges are the tiny arrector muscles in our skin. Like other mammals we have many thousands of them, each attached to a hair. In response to cold they pull the hairs up above the skin. In our more densely-furred relatives this creates an insulating air space which protects against heat loss. These muscles also pull the hairs up when mammals are involved in confrontations. As cornered alley cats show us when threatened, these arrector muscles pull the hairs up making the animal look larger and more imposing, thus intimidating adversaries and discouraging attack.

Arrector muscles still protect cats and most mammals against cold and threat but they do nothing now for us. We are Desmond Morris's "naked ape". Except for the tops of our heads and a few patches of hair which have been retained for functions such as sexual advertisement, nearly all of our hairy coat has been lost. The hairs left are so flimsy and widely spaced that no insulation or intimidation are possible now. Still if we come out of the shower into a cold room our arrector muscles tug at our hairs as they've done for millions of years, but we're made no warmer whatsoever. All we get for their efforts now are the tiny hills in our skin known as goosebumps.

Nor in threatening encounters do they nowadays make us appear larger and more intimidating. Still they 'come to our aid' when we think we hear an intruder late at night and when we are plummeting on a roller coaster, or reacting angrily to an umpire's decision. But scattered bristling hairs don't intimidate intruders or help us survive coaster rides or influence umpire's decisions. It's unlikely any human for thousands of generations has been saved from harm or even won an argument because of them.

Darwin knew about vestiges one hundred and fifty years ago. He realized that structures often become useless when circumstances change, and he knew that parts which have lost their usefulness do not therefore disappear. Evolution does not happen because of 'need'. It's not true that if you don't use it, you (or your descendents) will necessarily lose it. After all, we still have arrector muscles and many other vestiges after all these millions of years.

Why do vestiges persist?

One common reason vestiges persist is because environmental changes often happen far more rapidly than the processes which produce genetic adaptation. Organisms with long life spans, incidentally, probably have many more vestiges than organisms like insects or bacteria whose short generation times allow genetic adaptation to be hundreds or thousands of times faster.

There are other reasons besides time-lags which help explain why vestiges persist. Many continue to be produced, because, while they are not helpful, neither are they, on balance, significantly harmful, so natural selection cannot eliminate them.

Some vestiges, even harmful ones, persist because the genes which produce them also contribute to other more useful traits. Genes often have multiple (pleiotropic) influences so all costs and benefits must be weighed. If a gene contributes to structures A and B, as long as survival is helped more by A than it is hurt by B, B will continue to be produced indefinitely, riding on the coattails of A.

Another reason why some vestiges still exist though they

no longer serve their original function is because they have come to serve as precursors to other traits. Many embryonic structures like gill arches and gill slits, for example, have been retained because they became foundations for other, more recently evolved parts. Since jaws evolved from gill arches and Eustachian tubes are actually modified gill slits, jaws and Eustachian tubes cannot develop unless these ancient gill foundations are first produced during development. Sometimes things go wrong and the genes which produce the modern modifications mutate or are blocked and babies are born with gill arches and slits instead of jaws and Eustachian tubes. Some people explain these genetic tragedies and worse ones such as babies born with one head growing out of the other by saying God works in mysterious ways.

Vestigial behaviors

It's not something we often realize or think about much—you won't likely find even a paragraph in most biology or psychology texts—but it's true nonetheless—we humans and other animals have many vestigial behaviors for the same reasons we have vestigial structures.

Our domesticated animals provide many examples as we might logically predict since they live with us in the radically new environments recently created by our technology. Dogs and cats and cows and chickens for time-lag reasons alone are stuck with genes for many behaviors which were adaptive for their wild ancestors but which are useless leftovers now.

Dogs, for example, commonly walk around in a circle a couple of times before lying down to rest. Whether on the mud floors of Yanomamo huts or on nouveau plush in a New

York apartment, dogs repeat this ancient ritual several times a day. It is doubtful, though, that this behavior left over from wolves, helps a Pekinese survive in his Fifth Avenue pad.

Similarly, while most dogs are content to eat kibbles in the kitchen, when a bone or a large piece of meat is placed in their dish they usually take it *out* of the dish and carry it away to some other place in the house! One favorite place is under the dining room table. Not because of any sense of dining decorum, of course, but clearly because the table offers protection. Protection? From whom or what in a safe suburban house? It's ridiculous, but a pampered pooch, even if he is the only pet in the house and always has been, will take his prize under a table or bed and even growl ungraciously at the approach of his mistress who gave it to him in the first place, and has never shown any competitive interest in his food. (Rawhide, incidentally, and wooly gloves and socks, because of their animal chemicals, often release the same defensive hostility as many dog owners know.)

This irrational behavior is also caused by leftover genes from wolves. Like most hunting carnivores wolves are instinctively inclined to take the carcass of the prey, or a large piece of it to some defensible spot more free from competition.

Human vestigial behaviors

What about us humans? Do we have vestigial behaviors like our pets? Of course we do. Plenty of them. How could it be otherwise? We are animals after all, and like our dogs we now live in circumstances radically different from those on the plains of Africa where our behavior genes were selected. So we are stuck like they are with all sorts of ancient inclinations and instincts which don't make much sense in suburbia.

Consider our common negative reaction to spiders. Though infants do not have an instinctive aversion to spiders, we are biologically inclined to quickly learn this aversion during childhood maturation. Research studies have shown that we (and monkeys too) are predisposed to readily learn to react negatively to spiders (and other crawly critters), but not to flowers, for example. In any case this is a good example of a vestigial inclination which is deeply automatic and very difficult to control. How do most of us react if a spider runs across the covers when we're reading in bed, or more dramatically, runs up our leg? Violent flailing of arms, screams of horror, expressions of disgust and snacks spilled all over are likely. Even an arachnologist who *likes* spiders (at a safe academic distance, at least) and lives where there are no harmful species will react violently and risk his life if one drops onto his neck and crawls under his collar while he is driving on the freeway.

Death or serious injury from spiders is very rare in the U.S, but most of us still are instinctively averse (or instinctively inclined to learn this aversion) left over from long ago when tropical spiders did pose a serious threat. Automobiles kill about fifty thousand Americans each year. If our behaviors all made modern sense, people would scream and react violently every time a car approached, and visiting an automobile dealership would cause chills and aversive disgust.

Except when a spider lands suddenly on our body most of us can learn to overcome this natural aversion. We are much more rational and able to modify our behavior than even the smartest of our animal relatives thanks to our recently evolved cerebral cortex. In the last million years our cortex has grown more than twice as large as that of our ape relatives and first human ancestors. It has created our extraordinary powers of learning and logic and gives us potential for great behavioral flexibility and has allowed us to be somewhat free of the chains of our biological past.

But we're not transcendent yet. Not even close to flying free of our genes. As Harvard's great sociobiologist, E.O. Wilson, has indicated, even our cultural behaviors are on a "genetic leash". For one thing, our cerebrum, though marvelous, is only one part of our brain. It is by no means in full control of our behavior. There are many parts of our brain much older than the cerebral cortex which still exert profound (often vestigial) influences on our behaviors, as Paul MacLean described when he was director of the National Institute of Mental Health. The R-complex, for example, which first evolved in the reptiles, and the limbic system which appeared scores of millions of years ago in the early mammals still produce much the same aggressive and territorial and sexual behaviors in us today much as they do in our reptile and mammal relatives. Like all animals we humans have been extensively prewired by our evolutionary history. No theory of human behavior which ignores our ancient biological behaviors will ever explain much about us. It's no wonder that Freudian theory and a half a century of behavioral psychology failed so miserably to provide us with much self-knowledge.

Our cerebral cortex is rather like a young naïve boss of an old company whose nature and history he knows little about. A neophyte suddenly charged with control of an old enterprise of bewildering complexity. A company with a great diversity of workers, many of whom like the older parts of our brain are often old-fashioned and incompetent, carrying out archaic processes, unaware that times have changed, still producing antiquated products worthless in the modern market.

A company with many potential saboteurs on the lower floors not yet explored by the new boss. You can have ten thousand cerebrums at a football game and not one of them aware of the R-complex and limbic influences which brought them there and explain the behaviors they are watching.

Just as our arrector muscles still tug pointlessly at our hairs, the circuits and chemicals of the R-complex and limbic system and the genes which produce them still influence us today though the behaviors they promote are often vestigial now. They did not conveniently disappear when we acquired culture or because they have recently become useless or maladaptive. There hasn't been enough time. It was only yesterday in geological time that our ancestors were primitive hunters and gatherers on the African plains. And it's only been about 4 million years since our line diverged from that of our closest ape relatives. The genes which influence our behaviors today in our stadiums and churches and offices and singles bars are almost all the very same ones which prevailed because they enhanced survival long before we were even humans. We not only have Neanderthal brains in the nuclear age as David Barasch of the University of Washington has pointed out, we have primitive reptilian and mammalian brains as well.

Ironically, some of our most irrational vestigial predispositions are schizophrenically produced by the very same cerebral cortex which otherwise inclines us to be logical and rational. It's not just the R-complex and limbic system which are dangerously old-fashioned. Robert Trivers , author of *Social Evolution*, and James Gould of Princeton and many other scientists have pointed out that we have all sorts of cerebral leftovers too. They include natural inclinations to fabricate—that is to make up lies about the world, to say we know things which we don't, and to rationalize and justify, and even to subconsciously deceive ourselves so that we believe our own lies. All of these were useful behaviors in the past but not so much today in a world which is becoming increasingly scientific where our survival depends so much on knowing the truth about our behaviors. It's as though the young boss received constant misinformation and bad advice from other top administrators intent on preventing him from understanding the true state of company affairs. There are saboteurs at the highest levels in our corporate brains.

Many of our vestiges such as aversion to spiders and snakes are quite harmless and can sometimes even be useful. It's possible that the spider could give a nasty bite, and if you don't know which few snakes are poisonous it's probably better to avoid them all.

Some other vestiges cause us more trouble; for example our fondness for sugar as described in the chapter on *Instinctive Attractions and Aversions*. Sweetness was a good indicator of nutritious fruit for our ancestors, and so natural selection favored genes for liking sweetness. There was not a problem because there was no way our ancestors could damage their health by eating the limited amount of sugar found in fruits. Today though, now that we have refined sugar and it is cheaply available in very concentrated form in large amounts our once-adaptive fondness now causes many cases of tooth decay, diabetes, and obesity.

Another behavior which should be considered vestigial in modern societies is 'falling in love at fourteen' or fifteen or sixteen or seventeen. We call this 'having a crush.' We are biologically primed, partly by a strong sex drive in the teen years, to form pair bonds. Hopelessly, helplessly, head over heels, driven to reproduce, we fall in love more powerfully and irrationally than we ever will again. It's an instinctive evolved program which was adaptive for most of human history, and still works well in primitive societies. The ideal time for females to mate and start producing children was during the teen years. Women didn't live nearly as long then as now and didn't need to go to college. Those who delayed mating much longer than that were less likely to survive to raise their children and help the survival of their grandchildren. Natural selection, therefore, favored genes which strongly inclined women to form powerful bonds and mate during the teen years.

Our natural inclination to become neophobic as described in the chapter on *What's Fun and What's Done* should also probably be considered vestigial today. Curiosity killed the cat often enough among our earliest ancestors that neophobia was probably adaptive then, but with exponential change of technologies and the constant need to keep learning to keep up, inclinations to neophobia have become vestigial now.

Another behavioral inclination which was once adaptive but is usually vestigial today in modern societies is our irrational evolved respect for and deference toward taller men as described in the chapter *Tall Bishops and Genuflection Genes*.

Perhaps our most harmful vestigial behavior today is our easily stimulated own-groupist inclination to be hostile to other groups within our species who are different from us. The play form of this predisposition which is manifest in our enthusiasm for team sports is harmless enough, but this us-against-them aggressive predisposition which produces racism and war *(discussed in the chapter on Football, Patriotism, Women, and War)* given our technological capacity for enormous destruction and death is much more threatening to our future: our most dangerous vestige.

Our official scientific taxonomic description is Homo

◄ CHOOSY WOMEN AND CHEATING MEN

sapiens (meaning wise man). That's appropriate enough; we *are* the smartest species on earth by far. But because we are stuck with so many leftover primitive inclinations which are often maladaptive and even destructive in our modern environment, Homo *vestigius* might even be a better name for us.

CHAPTER **12**

Football, Patriotism, Women, and War
(Evolution and own-groupism)

GO INTO A bar where men are watching a football game on TV and pick out some guys who are very emotional, loudly cheering and jeering. When there is a commercial break, ask them "Why do most guys like football so much?" If you're a woman asking this question, the guys might roll their eyes and smirk as if to say "Typical dumb women-just-don't-understand question." If you're a man, even if you look like a regular guy, you may be regarded with suspicion. What kind of guy asks a question like that? They may think you are a communist or gay.

Whatever the answers they give, it will likely be the first time they've ever been asked this question and even if they have spent thousands of hours watching or participating in team sports, it's unlikely they have ever thought about *why*! You might as well ask why they like sex, or why they sleep, or why they scratch when they itch. We don't think much about why we behave as we do when our behaviors are influenced by our genetic inclinations.

Am I suggesting that we (men at least) have football genes? No, certainly not. Nor do we have genes for soccer or hockey, or any other team sport. What we do have though, is a genetic predisposition which inclines us to *own-groupism*; an instinctive us-versus-them predilection to favor our own group, and to compete aggressively, often with hostility, against other human groups. Football and other team sports are cultural inventions to be sure, but these cultural waves are just the surface above deep biological currents. **Team sports cannot be adequately understood** *without* **reference to biological behaviors shaped by our evolutionary past. Nor can we understand why we are patriotic, or why we wage war, without identifying the deep evolved biological inclinations of our species.**

Own-groupist, us-against-them behaviors have evolved in many social animals from bees and ants to wolves and lions, to chimps and humans. Natural selection favors cooperation and favoritism to members of one's own group for many reasons. Cooperative hunting and territorial defense (lions, wolves, chimps, humans) are among the most important. You help me, I help you, and we both benefit. This reciprocal altruism is the basis for much social cooperation.

Kin selection too (as described in chapter 8) is especially central to the evolution of own-groupism. When animals live in family groups or extended family groups and most of the individuals are genetically related, cooperative behaviors help the survival of one's genes in relatives' bodies. Just as caring for one's own young promotes survival of one's genes, so does defending one's brother against others.

As a general rule, the closer the genetic relationship, the greater the altruistic cooperation favored by natural selection. The colonial insects like worker ants and bees are more closely related to one another than most animals with the workers sharing 75% of the same genes. So it's no wonder that altruism toward one another is so great. Honeybees, for example, die when they sting in defense of the colony. The loss of one set of genes by this death results in the survival of many other sets of the same genes in their sisters so natural selection has favored this kamikaze altruism.

The flip side of altruistic cooperation with members of one's own group is instinctive antagonism toward other groups within ones species. The lions in the neighboring pride, the wolves in the neighboring pack, the chimps in the next valley are not close genetic relatives with whom it would pay to cooperate. Rather they are competitors who need the same resources, so natural selection has favored aggressive hostility toward them. Wolves and lions are never nice to outsiders from another pack or pride who wander into their territory, and they will often kill intruders, thus reducing competition. Chimpanzees will even invade the territories of other troops for the sole purpose of killing whomever they can find, usually other adult males but even babies and defenseless old individuals who offer no threat. One consequence of this is that they are able to increase their territory and reduce potential competition. Interestingly, **it is males, coalitions of males**, not females, who conduct these murderous raids.

We humans are also strongly inclined toward own-groupism because of our evolutionary history. Our primitive ancestors lived in fairly small extended-family groups (less than 100) where most individuals were closely related, so kin selection was an important factor in promoting cooperation towards one's own group and antagonism toward other groups of humans who were not close relatives. Because of cooperative hunting and territorial defense (almost exclusively male activities), selection also favored strong inclinations for male bonding and male coalitions to fight against these other groups of competitors. It's still true today that primitive societies are typically hostile to the tribes across the river. Many of them refer to themselves as humans but refer to their enemies as animals, or dehumanize them as evil inferior *others*. It's no wonder that aggressive conflict with their neighbors is more or less constant.

Because we humans are so much smarter than lions or chimps, and because we have developed weapons, some scientists have argued that this likely ratcheted up the pressure for especially strong human male-bonding, coalitional cooperation, and intergroup aggression. Some scientists think, moreover, that one reason why the human brain has grown so greatly in the last million years is because the increased intelligence allowed us to better compete aggressively against other humans.

All animals have evolved to be competitive except for clones with exactly the same genes. All humans too are competitive. So we play individual sports like tennis and golf, and we have spelling bees and figure skating competitions, and bridge games. But few things inspire such intense competitive emotion in most people as do football and other team sports. Two million people showed up to celebrate in downtown Chicago when the Blackhawks won the Stanley Cup. Two million! How many would show up if Chicago won the U.S. scrabble tournament? Over 100 million Americans watch the superbowl on TV! Even if you count babies and ballerinas and old ladies in nursing homes that's one out of every three people in the U.S.! Own-groupism is a powerful part of human nature.

A visitor from another planet would see that all over the

FOOTBALL, PATRIOTISM, WOMEN, AND WAR

world there are stadiums filled with thousands of fans screaming and shouting, cheering and jeering, and sometimes many of the men are shirtless. What would they think is going on? Surely not a scrabble tournament or a spelling bee. (Men without shirts, are, of course, displaying masculine aggression against the opposing team, and also displaying for the females).

Bold front-page headlines and extensive sports sections in our newspapers testify to our intense own-groupist interest in team sports. Team sports games often preempt all regular programs on television. Even a monthly bridge game with elderly players, men and women both, is likely to be delayed a half an hour if there is a major sports playoff on TV. Most bridge players are very serious about the game so it really says something about us when team sports can preempt bridge. Major sports competitions, in fact, often preempt just about all other activities. "No, our daughter can't get married on that Saturday; it's the national playoff between Kansas and Iowa!"

Again it's men who are usually the most interested. In fact team sports is one of the main things most men talk about. Go anywhere, from your local gym or the barber shop or even to the drugstore, and men will ask strangers "What do you think of State's chances?" or "Wasn't that great; Clayborn making the touchdown in the last 30 seconds?" The few guys who don't care much about football just answer "Yeah." They wouldn't dare say "Which state's chances for what?" or "Clayborn who?" It is assumed if you're a man you must be into team sports, and certainly the success of your local team.

Fewer women than men get as emotional about particular teams but many women do show own-groupist loyalty to their local sports teams. When introduced on TV shows like Jeopardy and asked where they are from, they often spontaneously offer unsolicited support for their local team! "I'm from

Detroit. Go Tigers!" they say. One woman, when the show was aired in Boston, was booed when she said she was originally from New York and she favored the Yankees. Jeopardy, of course, has nothing to do with team sports. It just goes to show how pervasive our own-groupism is.

We don't often realize what the strong emotions evoked by team sports reveal about our biological nature, but intelligent visitors from outer space would surely think it an important clue to our evolution. They would see that millions of men in diverse cultures become more intensely emotional when watching team sports than they do when observing most anything else. They loudly yell support for their teams and often aggressively insult the opposing team, not only when they are in stadiums, but even when they are watching by themselves in their family rooms. Guys don't yell and scream and shout insults when they are watching Wheel of Fortune or National Geographic specials or even golf tournaments. It's *team* sports—us against them—that stimulates aggressive own-groupist emotion.

If visitors from another planet watched our national football playoffs and baseball world series, they would see that hostile competitive feelings sometimes build up to such a frenzy that when the home team wins, some fans displace unspent aggression with violent dominance displays by fighting and smashing cars and such. Chimpanzees act much the same way in group-combat with adversaries. When victorious, they too celebrate with riotous aggressive group display. This happens because winning, whether it's a football game or a tennis match, or passing the bar exam, raises testosterone, potentially increasing aggression (and also interest in sex). One can only wonder how elevated testosterone might affect an old lady who wins the quilting competition at the State Fair.

FOOTBALL, PATRIOTISM, WOMEN, AND WAR

Losing lowers testosterone. It's better if you have lost a competition not to keep batting your head against the wall. Better to cool it for now and wait for another day.

Some people may object to the idea that interest in team sports has anything to do with biological causes because some men and many women show little interest. As if to say something cannot be biological unless it is universal. This is a common fallacy. Blue eyes and red hair are not universal, yet they are clearly produced by genes.

It's true that a few men don't like team sports much. In some cultures there isn't even any opportunity for our innate own-groupist inclinations to be expressed in this play form. In any case, there are many genetic reasons why some men don't like team sports. Many gay men (with female-type brains) don't like team sports as much as straight men. Many gay women (with male-type brains) are generally more interested in team sports than most straight women.

Also, brain sex aside, there is a lot of genetic variation in people's brains that has nothing to do with gender but which influence our likes and dislikes. Some completely straight men, for example, are very individualistic, and while they may be very enthused about golf or tennis, are less inclined to subjugate themselves to group authority, and are less interested in team competition. Just as some people are taller or shyer or more logical than others, some people are naturally more own-groupist than others.

Patriotism

If a boy doesn't like team sports and especially if he prefers girl games, he is likely to be teased and may be beaten up by other boys. Own-groupism again. You're not one of us! If he doesn't care much about team sports but goes out for track or plays golf or tennis, or engages instead in a sport or hobby which isn't 'feminine', it's usually no skin off anyone's nose, though, and he may be well-liked by other boys.

If a guy (or gal) is unpatriotic though, it's quite another thing. People are expected to show 'team spirit' and to support their home team or their local gang or city and, unless they are a repressed minority, their country. It may seem silly, but if a guy doesn't pledge allegiance at a rally, or fails to stand up and remove his cap at a football game when the national anthem is played, people will often shout at him and demand that he display his patriotism. **(Interesting that the national anthem is played at team sports games)** The reaction of fans at a game to someone who doesn't stand is rude, perhaps, but understandable. Few things incite group hostility more than failure to patriotically pledge allegiance to one's group.

Football, of course, is play behavior, albeit very serious play in the minds of many. Still, as long as you support your team, you can like it or not, take it or leave it, and it doesn't matter much to most of us. You can't take it or leave it though, when it comes to patriotism. We are all expected to be patriotic. Patriotism is, of course, another manifestation of own-groupism. It is usually expressed as unquestioning allegiance to one's group, especially one's country, and may even suppose a willingness to give one's life if necessary. Many super patriotic people won't even consider the possibility that their "enemies" are reasonable and good and might even be right. No, It's my country right or wrong! For them, blind allegiance automatically trumps discussion about who is right or wrong.

Patriotism is greatest during times of war, or when there is a threat to the country as happened with 9/11. At such times it encourages blind, even irrational support for leaders. 9/11 was the best thing that ever happened to muster support for George W. Bush and his administration and the invasion of Iraq. Politicians often exploit our patriotism by calling their opponents un-American or by inventing threats or inciting fears of enemies, real or imagined, as Joseph McCarthy did in the 50's, by labeling his adversaries as communists. Several commentators on T.V. (like Glenn Beck) appeal to primitive and paranoid us-versus-them own-groupist feelings in their viewers by labeling (libeling?) their liberal enemies as communists and Marxists or Muslims. Even though these charges are completely false they are effective in generating animosity and often hate.

Some extremely patriotic people believe that anything less than blind faith in our government or military campaigns is tantamount to treason. They can't accept that people who oppose a war, for example, may be among the best citizens. Mention to them that scores of thousands of innocent civilians were killed in the Iraq war and they think you are unpatriotic. Argue that the Patriot Act violates our constitutional rights, and you are unpatriotic. Mention that we were attacked by Saudis, not Iraquis, and you are unpatriotic. Many people have such strong us-versus-them inclinations that they even demonize our allies. They don't want to hear about the French healthcare system, and they are opposed to the United Nations, and they are strongly opposed to the idea that there will ever be a world government. My country right or wrong, us-against-them patriots they are.

Patriotic allegiance to one's own group was highly adaptive for millions of years of human evolution. It bonded people to cooperate and to defend against other groups, and it still promotes survival of the group and one's own genes in relatives today in primitive societies, but it doesn't make so much sense today in modern societies. Surely patriotism is not always a good thing. Hitler's youth were willing to sacrifice their lives to fight against us in WWII because they were biologically inclined to be patriotic. Arabs and Jews kill one another year after year because they are patriotically own-groupist.

War

Warlike behavior is organized murderous aggression of one group against another (usually of the same species) and it often evolves where there is intense competition for resources. Warring individuals do not have to understand why they are behaving in this way. Warlike behavior is produced by ancient instincts.

Ants certainly don't know why they wage war. Armies march automatically attacking neighboring colonies and causing vicious carnage. Soldier ants chop off heads and legs of rivals who are usually members of colonies *of their own species*, and slaves are often captured and forced to work in the colonies of the conquerors. All of this without conscious intent or awareness of why they are behaving this way.

It's significant that our closest animal relatives, chimpanzees, wage war too. They bond together first with mutual grooming and then march silently in a single file to invade the territories of other chimps and very brutally kill them for no apparent reason. It's significant too that it is males, coalitions of males who do this, and they obviously enjoy this brutality, screaming and jeering and cheering. Of course they don't know *why* they enjoy this any more than guys know why they

FOOTBALL, PATRIOTISM, WOMEN, AND WAR

enjoy team sports. Their biological inclination to kill other chimps who are not part of their group is instinctive.

The most dangerous and tragic consequence of own-groupism is human war. Football results in lots of injuries but rarely death. Wars though have killed hundreds of millions of people. More than 100 million were killed in just WWI and WWII. So many wars: the Greek and Roman Wars, the French-Indian War, the War of the Roses, the Spanish-American War, the War of 1812, and the war in Afghanistan. On and on; so many wars. Hutus and Tutsis, Arabs and Jews and so on. Considering that chimps practice warlike behavior, it seems likely that war has been almost continuous for at least as long as we have been humans.

Among the reasons why we humans *say* we have wars, are access to resources, disputes over territory, threats from others (real or not), religious conquest, and revenge. All sorts of justifications are given by those who start the wars. Of course, no rationale is needed for retaliation. **Some primitive societies don't even offer any rationalizations for war. It's reason enough to fight just because they are them, not us!**

Whatever the reasons or rationales we give for wars, evolutionary psychologists say they are not sufficient. To understand human war we need to understand why natural selection shaped the deep psychological mechanisms which incline men, like chimps, to readily form coalitions to aggress against other groups. It may not matter if the guys in the bar understand why they like football but it is important for us to understand why we (*men especially*) came to be so biologically prepared for war. Why it's mostly *boys* in societies all over the world who engage in war-play and show such natural enthusiasm for us-versus-them, smash-them, kill-them violent video games.

In the Newsweek story *Why Men Love War* (May 10 2010) Evan Thomas said "War has been for almost all peoples and all times the purest test of manhood." Chris Hedges, former war correspondent for the New York Times said, " It is a thrilling addiction and a wretched curse—a force that gives us meaning." That was certainly the case with the lead actor in *The Hurt Locker*. Sebastian Junger, author of *War*, has suggested that warfare can be addictive for men because of the strong feelings of group inclusion with other troops, a sense of purpose and identity, a certainty about their role and relationships and value to the group.

It is of interest here that of hundreds of societies studied, there is not one documented case of women forming aggressive coalitions to wage war. It's men's evolution we need to understand. Why were men's brains shaped so differently by natural selection? Clearly there must have been some reproductive advantage for men, but not women, to wage war. For hundreds of thousands of generations men who were genetically-psychologically inclined to be warlike *must* have had high reproductive success, or these inclinations expressed in males would not be so common today.

The main reason why warlike males had more children are known to evolutionary anthropologists but may be quite surprising to most readers. **It's this: men who waged wars for most of human history had more children, left more genes, because they had more wives and more extramarital copulations and therefore more children than men who were not inclined to be warlike.** The evidence for this is very strong from the study of preindustrial societies who today still live in circumstances like that of our primitive ancestors. To understand why we behave as we do today, we must understand the conditions we lived under for hundreds of

FOOTBALL, PATRIOTISM, WOMEN, AND WAR

thousands of generations when natural selection shaped our behaviors.

It turns out from studying many diverse preindustrial societies that **a major reason, why men evolved to be more warlike than women, was for reproductive access to women**! The Yanomamo of South America, a tribe courageously studied for many years by evolutionary anthropologist Napoleon Chagnon provides a good example. Yanomamo men are almost constantly at war with men from neighboring tribes. They form coalitions to stage raids and kill men from the other tribes, and they often capture women with whom they mate. Not only do the warriors have more children because of the captured women, they are held in high esteem by the women in their own tribe so they have more wives and extramarital copulations and about three times as many children than men who do not participate in war! By favoring the warriors as sex partners the women are shaping their sons, and males in general, to be warlike. **Because of female choice, women have largely designed men to be the way they are. So if our female ancestors behaved like Yanomamo women and preferred to mate with killer males, one could argue we should blame women for war.**

Even if the Yanomamo are not typical of primitive societies and presumably, our ancestors, studies of many preindustrial societies show that capture of women is a major cause of war in 45% of them, and access to gained resources because of war (which makes men more attractive to women) accounts for another 39% of the reasons for war. Directly, through capture of women, or indirectly because of gained resources which attract women, some 84% of wars in primitive societies result in increased reproductive success for the warriors. It seems that access to women is by far the most important unconscious motivator of war. **Attractive young women as cheerleaders at football games seem symbolically very significant!**

Just as men bond together to form aggressive coalitions in war, men in prison (but not women though) typically divide up into rival gangs revealing, again, the natural inclination for men to bond together for offense and defense.

Gangs of young men, so common in our larger cities are also testimony to the own-groupist inclinations easily stimulated in men. Not surprising. It's almost exclusively young men, not women, who form gangs to defend turf and resources (such as drug territories) and to bond aggressively against other gangs. And it's not surprising that the research shows gang members are more attractive to girls than non–gang members in the same neighborhoods: they have many more female sex partners than young men who are not members of gangs.

Warfare is still genetically adaptive for the tiny percent of the world's men who still live in primitive societies but it is not adaptive today for the great majority of men who live in modern industrialized societies. In fact it is vestigial. Men in modern societies who fight in wars do not gain access to more resources and (unless they rape) do not gain access to more females or have more children than men who do not risk life and limb. Still, just as most men like team sports without understanding the evolutionary reasons why, many men, even if they would be reluctant to give twenty dollars to a stranger in need, will often, without much thought or reservation, risk serious injury or death to serve their country in war. It doesn't even have to be one's own country. The male predisposition to fight other groups is so readily tapped that leaders (like Gaddafi) can easily enlist mercenaries from other countries who murder others without conviction or cause simply for a day's pay. This natural own-groupist inclination to readily participate in war is not only *not* adaptive in the modern world, it is extremely dangerous to all of us now. Our technology which increasingly produces ever more terrible weapons of mass destruction, capable of killing millions of people in a few minutes, threatens all life on earth. As David Barash of the University of Washington said "We have Neanderthal brains in the nuclear age."

There are many human phenomena other than team sports and patriotism and war which are the manifestation of our own-groupist inclinations. Xenophobia and homophobia, for example, stem from these roots. Gays, especially gay men are not only discriminated against but even hated, vilified, beat up and sometimes murdered (interestingly usually by other men) because they are them, not us. Own-groupism also generates religious intolerance, hostility to foreign aid, hostility toward immigrants, and the United Nations, and hostility to 'bleeding heart' liberals who suggest we should Christianically treat others like ourselves.

Political leaders often exploit our own-groupist inclinations by claiming that oursiders; them not us, are the cause of conflict. Even despotic rulers like Gaddafi of Libya and Bashar Assad of Syria, who killed many of their own citizens when they protested, claimed that the protestors were foreigners, part of a conspiracy.

Religious beliefs often provide fuel for violent own-groupism. Radical Muslim religious extremism is, of course, responsible for producing many violent terrorist attacks. But Christian fundamentalism too has often been associated with violence. Timothy McVeigh's Oklahoma City bombing of the Murrah building was significantly inspired by the influence of the Christian Identity movement. The Army of God, was responsible for not just the bombing of abortion clinics but shooting deaths as well. Speakers at their meetings have called for murder of abortion providers and have suggested that gays should be put to death. The Ku Klux Klan has publicly advocated white-supremacy and racism and anti-Semitism, and has carried out lynchings and other murders. This is violent terrorism performed by mostly Christian men who symbolically burn crosses at rallies. Readers interested in much more discussion about the violent own-groupism (and other evils) practiced by religious fundamentalists might want to read *The Fundamentals of Extremism* by Kimberly Blaker, *The End of Faith* by Sam Harris, and *The God Delusion* by Richard Dawkins

Racism is one of the ugliest and commonest manifestations of own-groupism. Though often disguised where it is socially disapproved, it is a natural inclination we have to learn to overcome. When conservatives and tea-partiers expressed opposition to Obama's health care reform and other proposals, they called him a communist, and a socialist. and a fascist, and suggested he was like Hitler with those moustaches on his pictures. "He's not one of us" was their own-groupist message, and racism was surely a big part of it. During the midterm election cycle of 2010 many television ads were deliberately designed to appeal to our easily-aroused racist inclinations. One of the most memorable was the ad approved by Sharron Angle of Nevada which depicted menacing-looking latinos coming across the border with the clear implication that they threatened American families.

Many politicians even appealed to our primitive own-groupist feelings by claiming that Obama was not an American citizen! In the summer of 2010 only 42% of Americans believed Obama was definitely born in the United

FOOTBALL, PATRIOTISM, WOMEN, AND WAR

States. 27% of Americans and 41% of Republicans thought he probably or definitely was *not* born in the U.S., and one in four Republicans thought he was the antichrist, and 31% of them said they thought he was a Muslim! By March of 2011 some 8 months later despite considerable press coverage that Obama was definitely born in the U.S. in Hawaii, 51% of Republicans, even more than before, said they believed he was born in Kenya! This preposterous belief gives credibility to Bill Maher's claim that most racists (in the U.S) are Republicans!

It's not hard to imagine why so many Americans would believe these patently untrue charges, and were so willing to ignore the facts. Many own-groupist people clearly feel that Obama is not 'one of us'. For some of them his Arabic name is enough to create hostility. Again, racist own-groupism is surely a big part of this. And, perhaps resentment as well, that Obama is one of the least own-groupist presidents we have had. He does not primitively divide the world into the good guys and the bad as George W. Bush did. In any case, nobody questioned Bush's citizenship or religion or painted Hitler moustaches on his picture when they disagreed with his policies.

Some people are much more inclined to own-groupism than others, and it seems likely this has a lot to do with genes. Behavior genetics studies have shown that most personality traits and even social attitudes and political opinions are strongly influenced by genes. Some research studies in recent years have shown, for example, that genes have a big influence on whether a person is conservative or liberal. Surely genes must play some role in influencing how much a person likes team sports, how patriotic they are and how willing they are to go to war without much question.

Our basic evolved inclination to be own-groupist is manifest in many different ways. It would be interesting to study whatever correlations may exist between enthusiasm for team sports, patriotism, racism, opposition to immigrants, homophobia, opposition to the U.N., and own-groupist political convictions, since these dispositions do, in fact, stem from the same genetic roots. Are football fans more likely to be extra patriotic? Are racists more inclined to be warlike? Do religious conservatives practice more own-groupist violence than atheists? We know from several recent studies that conservatives in general, are genetically different from liberals. Do conservatives and liberals differ genetically when it comes to own-groupism? There are lots of interesting possibilities for future research if we want to better understand ourselves.

CHAPTER **13**

Why Are People Gay?

THERE ARE HUNDREDS of millions of homosexual people in the world but it is often difficult to know who they are. One reason is, of course, because the hostility, the hatred, and the religious condemnation of gays is often so great, many people will not admit the truth even if they are assured the survey is strictly confidential. In several countries, in fact, if they are discovered, homosexuals face the risk of long imprisonment or the death penalty.

Another reason it is difficult to know how many people are gay is because there is confusion about how homosexuality should be defined. If homosexuals are defined as those who *have sex* with their same gender, the estimate is lower than if homosexuals are defined as people who are *physically attracted* to those of their same gender. Adding to the confusion, there are lots of people who are sexually attracted to their same gender but are married to (and have sex with) someone of the opposite gender not because of attraction but because of convention, or because of a love relationship, or because they want a family. These people should probably be described as gay or homosexual even if they do not self-identify as such on questionnaires. Sometimes they get off the homosexual hook by saying they are bisexual, though the evidence suggests that being physically attracted to both sexes is very rare especially for men.

Another problem with some surveys is that young people do not always *know* if they are gay or not. They may ignore embarrassing or guilty same-sex attractions or dismiss them as temporary or unimportant. Many gay teens and young adults date members of the opposite sex, supposing that normal heterosexual attraction will eventually come naturally. They often report that they did not realize they were gay until they were well into their twenties so they would not have identified themselves as such if they had been surveyed earlier.

Despite the difficulty of determining exact numbers of gays, many surveys and scientific studies have been done so it's possible to make some good estimates. If we average the estimates from those studies scientists consider to be the most reliable, the estimated percent of homosexuals in most populations is about 5%, or one in twenty individuals. By this calculation more than 15 million Americans are gay, as are 75 million Chinese, 67 million Indians, and 25 million Europeans. Seven to eight million Nigerians are gay and therefore risk imprisonment for up to 14 years. In areas of Nigeria with Sharia law, homosexuals have been executed, and a person can get up to four years in prison just for advocating gay rights. Pakistan is more lenient. Its 10 million gays face the prospect of no more than two years in prison. In Uganda, where homosexuality is illegal, a proposed bill inspired by American evangelical Christians in 2009 would provide the death penalty! Calculating how many homosexuals there are by religious affiliation rather than by country, two of the religions most opposed to homosexuality are evangelical Christians despite the fact that about 20 million of them are gay, and Muslims (about 75 million Muslims are gay).

What are the proximate causes?

One thing scientists know a lot about is the mechanisms which cause a person to be gay or straight. For at least 40 years many research studies on sexual orientation have shown is that **it is the influence of hormones during prenatal development which determines whether a person will be gay or straight.** The human brain, like that of typical vertebrates, is programmed by default to be female unless sufficient testosterone (produced by a male fetus) is present during the second month of prenatal life to convert it into a male brain. If there is not enough testosterone for whatever reason, even if the fetus is otherwise physically a male, his brain will remain female, his behaviors as a child will usually be more feminine than masculine, and because he has a female brain, as an adult he will be sexually attracted to men. Similarly if the body of the fetus is female but there is too much testosterone for any reason, her brain will become masculinized and she will usually show male-typical behaviors as a child, and as an adult, she will be sexually attracted to females. **Homosexuals are people who have a brain gender which is different from their body gender. Because brain sex strongly influences behaviors and self-identity, most gay men should probably be thought of as women who were born with a man's body, and most lesbians should probably be thought of as men who were born with a woman's body.**

Characterizing homosexuals as people who have the 'wrong' body which doesn't match their brain is reasonable but it is often too simplistic because many people do not have one-size-fits-all standard male or female brains. There is a lot of variation, for example, in *how* masculine or feminine the brain becomes. There are some lesbians whose behaviors are more masculine than most men, and there are some gay males who are more feminine than most women.

There are also some gay men who are very masculine in their behaviors but are erotically attracted to men, and some so-called lipstick lesbians who are very feminine but are attracted to women. Sometimes this is because gay people feel shamed, and because of strong social pressure, they learn to behave in gender-appropriate ways. In some cases, though, the evidence suggests this is because there are two separate centers in the hypothalamus, one of which influences gender behaviors while the other influences erotic attraction.

These centers are usually programmed the same way so that gender behaviors and erotic attraction match so that they are both *either* masculine *or* feminine. Sometimes, though, one center remains feminine during fetal development, but the other is masculinized in which case the person shows the typical gender behaviors of one sex but the erotic interests of the other. So this, for example, can produce a very masculine feeling and acting man who is sexually attracted to men, or a very feminine woman who is sexually attracted to women.

Here is some of the evidence which shows that sexual orientation is biologically determined and not a choice:
1. Homosexual behavior is common in vertebrates, especially in the mammals, and in our primate relatives, and homosexuality has been reported in all human societies from biblical times until present; from American Indians to Germans, from Zulus to Spaniards, from ancient Aztecs to present day Canadians. And, the

percentage of gay people (around 5%) is about the same in all cultures. When a trait occurs in many species and in hundreds of diverse human cultures over time at about the same frequency, this very strongly suggests biological causation.

2. When homosexuals are asked how they came to be gay, virtually all say that they believe that they were born that way and that they certainly did *not* choose to be gay. Most of them report that as children they felt like the opposite sex for as long as they can remember, and in fact, many research studies have confirmed that most gay males, when they were children as young as two years old, showed gender-atypical behaviors, behaving more like girls, and that gay women as children typically felt and behaved more like boys and preferred boy's activities and interests. Both gay males and females also report that their first sexual attractions were toward members of their own gender. The sense from early childhood, many gays report, of being born with the wrong body is so strong that increasingly gays are choosing reconstructive surgery at great financial cost to make their bodies match their brains.

3. Homosexuality can easily be created in experimental animals by manipulating prenatal hormones and thus changing their brain sex. Female rats and monkeys, for example, if given too much testosterone during prenatal life develop male brains and later behave like males and are sexually interested in females when they are adults. Male rats and monkeys deprived experimentally of sufficient testosterone during brain

development behave like females as adults and try to mate with other males.

4. Many studies have shown that in humans, just as in other animals, if a female fetus receives too much testosterone during prenatal life, as is the case with girls with congenital adrenal hyperplasia (CAH) the girl, so affected, will almost always show male-typical behaviors as a child and will usually show sexual interest in females as an adult. Likewise, human male fetuses deprived of enough testosterone during early prenatal life are born with male bodies but with female brains and they typically, just like the rats and monkeys, show female-typical behaviors when they are young boys, and when adult, they are attracted to men, not women.

5. Research has also proven that there are several anatomical differences in heterosexual and homosexual brains. The region in the hypothalamus, for example, which influences sexual orientation and gender behaviors is, in gay men and straight women, half the size of that region in straight men. Another difference discovered by researchers at the Karolinska Institute in Stockholm Sweden, is that the right hemispheres of straight men and gay women are slightly larger than those of straight women and gay men. PET scan imaging also revealed that responses to emotional stimuli by the amygdala (involved with emotion) of the brain by homosexual men was much more like that of women than like that of straight men.

6. There is no scientific evidence that sexual orientation can be chosen by animals or humans. Sexual attraction is always biologically determined, always instinctive, and never a matter of choice in any animal. A lion or a monkey cannot choose whether to be sexually attracted to males or females, any more than it can choose the color of its fur. Nor is there any evidence that humans can decide whether to be gay or straight any more than they can decide the shape of their noses. There is, moreover, a great deal of scientific evidence that sexual orientation can not be changed. Despite the claims of some fundamental Christian ministries that gays can be made straight, there is no scientific support for such claims and there is considerable evidence, according to the American Psychological Association, that these programs do not work. The APA has also indicated that therapy meant to 'cure' homosexuality is not only ineffective but can be extremely harmful to the mental health of people who have same-sex attraction.

7. It is also clear that basic gender behaviors and sexual orientation cannot be changed by how a child is raised. Boys, for example, who were raised as girls (in one case because the boy's penis was cut off as a result of a botched circumcision) still grow up feeling like boys and are later sexually attracted to girls. Pink dresses and dolls and being raised as a girl from an early age cannot change a male brain and behavior and sexual orientation to a female pattern. Likewise girls have sometimes been raised as boys but no amount of trucks and footballs and encouragement to

be tough can change the brain sex, so these girls still feel and act like girls and as adults they are sexually attracted to men not women.

8. Studies of identical twins, fraternal twins, genetic siblings and adopted kids raised together and apart all prove that genes play a very important part in both males and females in determining whether a person is heterosexual or homosexual. If one identical twin is gay the other twin has a fifty percent chance of also being gay even if they are separated at birth and raised in different homes. (If genes were not involved the chance that the other twin would also be gay is only about 5%) The fact that the odds of both twins being gay are less than 100% shows that while genes are involved, some environmental factor must be involved as well. Something, probably in the prenatal environment, is activating the genes for homosexuality in one twin or suppressing it in one of them. Other genetic studies show that homosexuality runs in families; brothers, even if they are raised in different families are four times more likely to be gay. In one form of homosexuality the genes involved are located on the X chromosome inherited from the mother. In this type, just as one would expect, male relatives on the mother's side (her brothers and father and grandfather) are far more often gay than chance would predict.

9. One cause of homosexuality which probably accounts for no more than one third of the cases has been confirmed by several studies which show that the more older biological brothers (but not sisters) a male has,

the greater the chance that he will be homosexual. This seems to be due to an immune response by the mother to previous male fetuses. It is speculated that this may interfere with the masculinization of the brains of later born males.

Homosexuality can be caused by tumors or drugs or immune responses or even stress during pregnancy (according to some studies), or by anything that causes an unusual increase or decrease of testosterone during fetal life when the brain sex of the fetus is being determined, but these factors probably account for only a small percent of cases of homosexuality. Most homosexuality appears to be caused by inherited genes (which influence prenatal hormone levels).

So the evidence is overwhelming that in animals and humans; homosexuality is biologically determined during prenatal development, that testosterone level is part of the proximate explanation, that genes are usually involved, and choice has nothing to do with it. Because homosexuals usually show gender-atypical behaviors by the time they are two or three years old, the claim that homosexuality is a choice would require the ridiculous premise that toddlers are choosing to be gay.

Homosexuality and Evolution

Because homosexuality occurs in all human societies, because it is usually caused by genes, and because homosexuals are much less likely to have children, it is puzzling to evolutionary scientists why these genes are so common. Surely natural selection must have favored these genes somehow, but how could selection favor genes which make the individual less likely to reproduce? There are several theories about this but little proof so far that any of them is correct. One thing is clear though. Genes which cause homosexuality *and nothing else* could not be favored by natural selection. **Genes for homosexuality must produce some reproductive advantage for relatives of homosexuals or they would become very rare or extinct.**

Some of the theories are based upon knowledge of kin selection. As described in the chapter *Why Charity Begins at Home*, genes can often be favored by selection even though the individual does not reproduce, as long as that individual is helping the survival and reproduction of relatives who carry copies of those genes, as is the case with helper animals and menopausal women. There is no evidence that gay people today in San Francisco or New York significantly help the survival and reproduction of their close relatives, but perhaps for most of human history when we lived in small family groups. having a gay brother or sister or uncle or aunt who didn't reproduce directly but baby sat and provided resources and protection could have helped their relative's survival enough that the genes for homosexuality could be passed on by those relatives.

On the island of Samoa today homosexual men are thought of as a separate gender called fa'afafine and researchers from the University of Lethbridge in Canada have documented that these men are far more attentive and helpful to their own nephews and nieces than straight men are. But there has been no study of whether or not this extra parenting has increased the survival and reproduction of those relatives. Kin selection may help explain the evolution of homosexuality but so far there is no proof that gay people are like helper

WHY ARE PEOPLE GAY?

animals or menopausal women. And, the kin selection theories do not explain why helpers are attracted to their same sex, as opposed to being asexual.

One interesting fact which was revealed by the Samoan study is that the mothers of the homosexual fa'afafine had more children than most Samoan women and this supports a very different theory about the evolution of homosexuality; that genes for homosexuality are favored by selection despite the fact that they reduce reproduction by affected individuals because they increase reproduction by female relatives. A similar conclusion was reached in a large Italian research study conducted by Ciani, Zanzotto and Cermelli and published in PLoS ONE. They reported, just as in the Samoan study, that female relatives of male homosexuals were more fertile! More fertile? Well, they did have more children. It could be, though, that the genes, rather than increasing fertility per se, simply made these women more interested in sex and that's why they had more children. Maybe the genes involved cause females (and their male homosexual relatives) to be specifically attracted to sex with men. That would explain both the increased 'fertility' of these women and the homosexuality of the male relatives but so far nobody knows exactly how the genes work, so more research is needed.

Finally, there are theories which propose that there are several genes involved and they shift the brain sex during fetal life in a more female direction. If a male gets enough of these he will be gay and thus less likely to reproduce, but if he inherits only one or a few of these genes his brain sex will be shifted just enough in the female direction that instead of being supermasculine he will be gentler, more empathic, more socially sensitive and less aggressive and therefore more attractive to females as a better partner and potential father.

◄ CHOOSY WOMEN AND CHEATING MEN

Perhaps as interesting as the question why natural selection has favored genes for homosexuality is the question about why there is so much disgust and criticism and even hatred of homosexuals, especially of gay males by straight males. Gay men tend to be more sensitive, and less aggressive physically. They have male bodies but female brains and so they usually behave in a more feminine way than most men do. What's so terrible about that? Why should they should be despised and sometimes beaten or killed by some straight men? The chapter on *Football, Patriotism, Women, and War* which describes our own-groupist hostility to those who are different from us may help in understanding this.

CHAPTER **14**

Unintelligent Design

OPPONENTS OF EVOLUTION in the U.S., most of them fundamental creationists, have claimed that humans and all the earth's creatures are so marvelously complex that they could not have been produced by a natural evolutionary process, and must instead, have been created by an intelligent Designer. They have argued, moreover, that this concept, called Intelligent Design, (ID) should be taught along side evolution in high school biology classes. About half of Americans (including George W. Bush) have indicated they agree. Seems fair enough, they say. Present both sides.

Well, the scientists do not agree. Intelligent Design is not science and there is no scientific evidence for it. Not a bit. It would be like presenting both sides of the earth is round-versus-flat controversy or like teaching astrology in astronomy classes. The American Association for the Advancement of Science, the National Association of Biology Teachers, and the National Center for Science Education have all opposed the teaching of ID, along with evolution, because it is not science. They have taken much the same position as the National Academy of Science which has stated that: "Creationism, intelligent design, and other claims of supernatural intervention in the origin of life or species, are not science because they are not testable by the methods of science."

Despite the opposition by scientists to including ID or other forms of creationism in public schools, there have been several attempts by parents and school boards to require the teaching of these religious, non-scientific explanations anyway, so this has led to court battles. When challenged in court these attempts have, again and again, been soundly defeated. Judges have ruled that teaching ID and other forms of creationism in public schools violates the Establishment clause of the First Amendment of the U.S. Constitution. In other words, teaching religion alongside evolution in biology class is clearly unconstitutional. In a Dover, Pennsylvania case in 2005, a Republican, church-going, federal judge issued a statement of more than 130 pages taking this position against ID in the classroom and he sharply criticized the school board members who had put it in the curriculum. In the next school board election all 8 members of the board who had required ID in the biology class were voted out of office by the citizens of Dover.

It's interesting that the advocates of teaching intelligent design and other religious 'theories' in public schools in the U.S. are almost always fundamentalist Christians, and want only the Christian version taught. Not the Jewish version, not the Buddhist account of creation, not the Native American Indian stories, and certainly not the Islamic version from the Koran.

Even if it were constitutionally ok to teach ID in public schools, along with evolution, many scientists and philosophers of science have argued it would be a very bad idea for two reasons: First of all there is massive evidence for evolu-

UNINTELLIGENT DESIGN

tion, but there is absolutely no scientific evidence for ID. You might as well teach that the earth was created by invisible pink aliens from Mars. A second reason is that ID does not explain anything at all! If we assume that life is so complex that there can be no natural explanation, and therefore there must have been an intelligent designer, well then, who or what created the intelligent designer? This is a philosophically worthless argument which requires endless recursion. Base your philosophy term paper on this premise and you would flunk.

There is another, much more important reason, though, why ID is a very dumb idea: its basic premise is clearly, absolutely wrong! ID asserts that the design of humans and other creatures is intelligent. But nothing could be farther from the truth! We humans and other organisms are loaded with many features which demonstrate very bad, inefficient, and even hazardous design.

Consider our human backbone, for example. Made of exactly the same vertebrae as our animal relatives who walk horizontally on four legs, it is a very bad design for walking upright. About half of all humans sooner or later suffer painful and debilitating problems such as pinched nerves, slipped discs, hernias, and muscle spasms because the design for upright walking is so poor. Evolution by natural selection, of course cannot produce perfection since there is no planning. Everything is make-do, built, not from a plan and not from scratch, but upon what came before, so it's to be expected that our backs would give us problems now that we're so recently upright after many millions of years of horizontal locomotion. Imperfect design is exactly what you'd expect from the process of natural selection. But you'd think if our backbone were created by an all-powerful god-designer he could have done a lot better than to take the same old design of horizontal vertebrates and turn it vertical for us his special creature. Created in the image of God? Does God have a vertebrate backbone?

Another example of bad design is the unique placement in humans of the openings to the esophagus and the windpipe right next to one another in the throat. Biologists say this happened because of the evolution of speech. Other animals do not have the problem of food getting into the windpipe but humans do, and thousands of people (mostly children) die every year of choking because of this unintelligent evolved design.

The rather narrow opening between the pelvic bones is another example of bad design. Human babies have very big heads and it often happened before caesarian deliveries were possible that women and their babies died trying to give birth.

Not very good planning either for men's inguinal canals. Mammals, unlike most vertebrates which have their testes inside the abdomen, hold their testes outside the abdomen in scrotal sacs, so the testes must descend, around the time of birth, from the abdomen through the inguinal canal to reach the scrotal sacs. This leaves weak spots in the abdominal wall, not a problem for most mammals but in us upright humans, gravity sometimes presses the intestines against the openings to the inguinal canals causing them to widen, resulting often in the intestines becoming strangulated and gangrenous. Before modern surgery to correct this design flaw, death was often the result.

Vestiges (described in chapter 11) provide other examples of bad design.. Flightless birds like ostriches with their useless wings. Hip and leg bones in whales and boa constrictors. Five separate toes in horse embryos. Human embryos with gill

UNINTELLIGENT DESIGN

slits, fish-type kidneys and tails; all of this you would expect as a result of evolution. The fossil record shows that the ancestors of whales walked on land and the ancestors of snakes also had legs, and the first horses had five toes. And since land vertebrates evolved from fish, the gill slits and gill arches and fish-type kidneys found in human embryos are testimony of our evolutionary past. None of this makes any sense, though, if there was an intelligent designer. Why would an intelligent designer give whales leg bones and give humans embryonic gill structures and tails?

About two thirds of all organisms on this earth are parasites which torture and often kill their animal hosts. Many of these parasites are specifically 'designed' to attack humans only. Thousands of viruses and bacteria and protozoans and worms cause all sorts of horrible diseases and often agonizing death. Malaria alone kills about 2 million people each year. Sleeping sickness, river blindness, liver flukes, dysenteries, intestinal worms, and many other parasitic diseases cause terrible pain and suffering and often death each year to hundreds of millions of people, many of them innocent children. It makes sense that parasites would evolve by an unintelligent purposeless natural evolutionary process which has no foresight and doesn't care about the suffering and deaths of children. But it seems incomprehensible why an omnipotent God would create a world with more parasites than free-living organisms many of which cause horrible suffering and death. If you were God would you? What sort of an intelligent designer would create parasites?

Critics of intelligent design, many of them scientists, argue that if there were a designer who created life on this earth, he could not possibly be intelligent or competent by any reasonable definition of those words, given all the bad design we see everywhere. And, given the great number of parasites which cause such misery and death, he must be very cruel as well as incompetent. Many scientists and philosophers have taken the same position as Richard Dawkins on this subject; It is not even theoretically possible that there could have been an intelligent designer.

CHAPTER **15**

Why Are People Religious?

MANY ATHEISTS AND other non-believers are often genuinely puzzled as to why anyone would believe in God. And they find it hard to understand why so many people believe religious claims which seem absurd to them. They argue that these beliefs are irrational and illogical, and are not supported by credible evidence, and are, in fact, absolutely contradicted by scientific fact and even common sense. How can an intelligent person, they say, believe in gods and devils, and angels and miracles and prayer? How can someone believe the earth is six thousand years old, that evolution never happened, and that there is a heaven and a hell when there is not the slightest bit of evidence for any of these beliefs or even that God exists? How, they ask, can someone believe in the literal truth of the Bible or the Koran when both of these books are full of contradictions and represent the primitive pre-scientific myths of people thousands of years ago? Why would a rational adult believe any of this?

Questions of truth or lack of same aside, critics often wonder how believers can accept the inhumanity and violence so often encouraged and justified by religion. From the murders of heretics during the Christian crusades to the burning of witches, to the Ku Klux Klan, to genital mutilation of women in the Muslim world and the proposed death penalty for homosexuals (Uganda, 2011) tacitly accepted and encouraged by American Evangelical ministers. From the biblical injunction to kill people who work on the Sabbath, to the bombings in Bali and Mombai. From God commanding Moses to kill his son, to 9/11 and the many other murders of innocents inspired by religious terrorists.

The evolutionary reasons why our primitive ancestors were religious and why so many people still are today will be discussed later in this chapter but **the most important proximate reason why people are religious is that children are indoctrinated by adults** (parents, priests, rabbis, mullahs, etc.) This works because children have evolved to be susceptible to believe what parents and trusted adults tell them, especially when the tone is solemn. That's how children, born without knowledge of the world, have always survived. It doesn't matter if what they are told is irrational and illogical and is contradicted by science. We are biologically programmed to believe our parents when we are young. It's instinctual.

If a child could be raised without any indoctrination he or she would be free, in theory, as an adult to weigh all the evidence and consider alternative points of view about the nature of the world, and whether or not gods exist. But that doesn't happen. Children are almost invariably indoctrinated. They have little choice except to accept what they are told. It's a bit like imprinting in that the influence often seems to be permanently stamped on the child, and he usually retains the same beliefs, more or less, when he becomes an adult.

Jews almost never convert to Mormonism, Catholics

rarely become Buddhists, and it's almost inconceivable that an Episcopalian would join the Holy Rollers, or a Unitarian would become a Jehovah's Witness. When those nice ladies come to our front door to proselytize and hand out leaflets, it's very unlikely that anyone would say "You ladies make so much sense that I've decided to switch from being a Lutheran to becoming a Jehovah's Witness."

Sometimes we reject our childhood beliefs but we usually keep the religion we were raised with or join one which is very similar. Indoctrinated for years that the religion of our childhood is superior, or told (as is often the case) that it's the *only* true one approved by God, and sometimes even told that non-believers and people of other faiths cannot enter heaven, or may be punished in hell, it is difficult for many people to imagine significantly changing their religions. **Because the early indoctrination is so powerful and so lasting, and because children are not able to make reasonable decisions about what they should believe, and instead are encouraged to accept authority and ignore scientific evidence and believe on the basis of 'faith', several critics of religion have suggested that religious indoctrination of children constitutes serious child abuse.**

If a parent really wanted to respect his child's self-actualization and freedom to choose, and didn't want to be criticized for brainwashing, he *could* expose his children to many different religions and to the counsel of non-believers as well. This week the child goes to a Baptist Church, next week to a Methodist Sunday School, the third week to a Synagogue lesson for kids, and then in future weeks to a Buddhist temple, a Christian Science lecture and to atheist and secular humanist meetings as well. Of course the parent would have to be careful not to tell the kid his own beliefs or otherwise bias him for or against any of these religions and groups. It's hard to imagine, though, that many parents would think this is a good idea. It's clear that most parents don't want their children to be free to choose their religious beliefs.

For almost all of human history; for all those hundreds of thousands of generations when our ancestors were primitive hunters and gatherers, it seems likely that *everyone* was religious. Before scientific explanation was possible there was no alternative to believing in gods and spirits. How else to explain human existence and the nature of the world. What are the moon and stars? What causes storms and natural disasters? Why do people die, and what happens after death? We humans have always *needed* answers to make sense of our world. Our need for meaning is so great that lacking good answers, psychologists say, we make them up. This fabrication is a natural function of the human brain.

It's not surprising before scientific explanations were possible that people assumed that thunder and eclipses and droughts and diseases and mysterious deaths were caused by something like a powerful being who, like humans, was capable of desires and emotions and states of mind ranging from benevolence to wrath. It's no wonder as Aristotle and many philosophers since have said, that people created gods in the image of man. Without knowledge of physical causes of events, but knowing that people are causative agents it would have been quite natural to assume that powerful beings, rather like people controlled these things, and it would have been reasonable (and very comforting) to appease those gods by prayer and ritual and thereby influence the god's decisions. The inclination to believe in gods would have been natural and predictable. A major function of all the world's religions, even today is to explain why things

WHY ARE PEOPLE RELIGIOUS?

happen and how to make things better; to gain control by prayer and appeasement.

The most and least religious people

The most religious people in the world today are those who still live in preindustrial, preliterate, primitive tribal societies like our ancestors did for so long. And, the very most religious of them, are the ones who have little or no contact with the outside world, and no knowledge of modern scientific explanations. You won't find any skeptics, much less atheists, among the Machiguenga or the Hadza.

Of the people who live today in more modern industrialized societies, the most religious are Africans and people of the Middle East, Malaysia and Indonesia, (who are mostly Muslims), and after that, the countries of Central and South America (but not Argentina). The United States is nearly as religious with 60% of Americans saying religion is very important in their lives, and about 90% indicating they believe in God.

Canadians and Western Europeans are much less religious than Americans. Only 30% of Canadians (half as many as Americans) consider religion to be very important in their lives. For Germans that figure is only 21% and for the French it's only 11%.

The least religious people are Scandinavians, Estonians, and people of the Czech Republic. Only about 1 in 6 Swedes believe there is a God, and about 5 in 6 consider themselves skeptics or atheists or agnostics. While about one third of Americans believe the Bible is the actual, literal word of God, only 3% of Swedes think so. Among Americans, Republicans are much more likely than Democrats to believe in the literal truth of the bible (77% vs.59%).

What might account for such big differences in religiosity from one country to another? There are several variables other than indoctrination which help somewhat in understanding these differences, but none of them is sufficient to explain more than part of the variation and none of them explains why people came to be religious in the first place, but these correlations offer some insight as to why, in addition to indoctrination, people are religious

Income level Several studies including the Pew Global Attitudes Project (2007) have shown that religiosity and income level are inversely related in most countries. In other words, people in the poorest countries (most of them in Africa) are the most religious, and people in the richest countries including Canada and Western Europe are the least religious. There are some anomalies though; some oil-rich countries of the middle east, such as Kuwait, are quite wealthy but are also extremely religious (Muslim), and though the U.S. is the wealthiest of all large nations, it is far more religious than Canada or Western Europe. This is likely due, in part, to the fact that the distribution of wealth in the U.S. is much more uneven, with relatively few people owning more of the wealth, so there are proportionally far more poor people in the U.S. than in Canada or Western Europe. In any case, poor people in the U.S. tend to be much more religious than the wealthy. The high religiosity of the U.S. might also be partly due to the fact that most of the original settlers were super-religious (The original colonies were intolerant of non-believers, and in many ways were rather like theocracies).

Educational level How much education one has is a better predictor of adult religiosity than income level, in part because poor people generally have less education. In any case, many studies have shown that the more education one has, the less religious one tends to be. People with college degrees are usually much less religious than people whose education stopped with high school, and people with doctoral degrees are generally less religious than those with bachelor's degrees. The least religious of all are those with Ph.D's in science. It is contrary to science to believe anything on the basis of faith, where there is no evidence or logical rationale. It is a fact as well that the more distinguished a scientist is, the less religious he usually is. Some leading scientists, more often chemists or physicists than biologists, believe in something bigger and more powerful out there but among the most eminent of scientists such as Nobel Prize winners, it is extremely rare to find one who believes, like most people do, in a personal God.

Social Health and Security Income level and educational attainment, as indicated above, correlate closely with how religious people are, but there is a more important general predictor of societal religiosity; it is the overall social and physical health and security of the society. Gregory S. Paul has stressed this point well in a carefully-researched article: *The Big Religion Questions Finally Solved* (Free Inquiry Dec. 2008/Jan. 2009) So does Phil Zuckerman describe this relationship in his important book *Society Without God* (2008). Those societies which provide universal health care and social support programs and have low income inequality where few people are poor, and have low crime rates and high political stability (Western Europe and Canada) are the least religious.

It is suggested that this is because people are well cared for and socially and economically secure so there is less anxiety and fear and less need to pray for supernatural help. No need to appeal to God in Sweden or France or Canada, for example, to prevent a serious illness from bankrupting your family.

Intelligence Some people with high IQ's are very religious and some who are not very bright are not religious, but numerous research studies have shown that in general, intelligence and religiosity are inversely related. In other words people with high IQ's tend to be more skeptical and less religious than people of average or low intelligence. Mensa members, for example, are usually much less religious than people with average intelligence. A 2008 study by Helmuth Nyborg published in the scientific journal *Intelligence* found that Atheists had IQ scores 3,82 points higher than believers in liberal religions and almost 6 points higher than fundamentalists. Average IQ scores have been calculated for many different countries and when these are compared, again the data show that nations with the highest average IQ's, Japan, Korea, Taiwan, and Western European countries are the least religious.

There are many other factors which cast light on why some people are religious and some are not, but, again, none of them are sufficient to fully explain religiosity.

Genes One factor which is extremely important, probably at least as important as upbringing or intelligence or educational level in explaining why adults are religious, is what genes they inherit! Yes, genes! The evidence is very strong that genes make a big difference in whether people turn out to be religious or not.

WHY ARE PEOPLE RELIGIOUS?

Most people are very surprised when they first hear that genes play a large role in religiosity and they often express skepticism that this could be true. Surely, they say, people are religious because of what they are taught and what they experience. It seems preposterous to people who don't know about the genetic evidence that inclinations toward religiosity could be inherited.

But it is true. It has been scientifically proven again and again that genes play a major role in determining whether *adults* turn out to be religious. In fact behavior genetics research indicates that about 50% of the variation in people's attitudes about religion, ranging from extreme atheistic skepticism to unquestioning fundamental hyper-religiosity, is due to what genes they inherit! Genes may play a bigger or smaller role in particular individuals, but assessing the whole population, the research shows that about half of the variation in religiosity is due to environmental influences such as upbringing, and half is due to genes.

Two of the largest behavior genetics studies of religiosity were done by Virginia Commonwealth University and the University of Minnesota. Siblings, fraternal twins, and identical twins raised together were compared with those raised apart in order to determine the relative influences of genes and environment, and the results were clear that genes account for about half of the variation in people's attitudes and beliefs about religion. Identical twins, even if they were raised apart from birth and had never met were extremely similar, much more so than fraternal twins, in their religiosity or lack of same. The genetic influence is so strong that if a person is genetically predisposed to be non-religious, even if he goes to Catholic school for twelve years, goes to church every Sunday and serves as an altar boy, it often happens that all that indoctrination does not 'take.' It's like water off a duck's back. As an adult he is likely to turn out to be indifferent to religion or become agnostic or atheist. You can't make a blue-eyed person have brown eyes no matter what you teach them and while you can make a child religious by indoctrination, you can't make an adult religious if he is genetically inclined not to be. Adults who forsake religion after a religious childhood are typically those who are genetically predisposed to not be religious.

How do the genes work which incline us to be religious? Are there specific circuits and programs in the brain? Is there a *God Part of the Brain* as the title of Matthew Alper's recent book suggests? Not much is known about the location of such circuits, but that is not surprising since little is known about specific locations for most behaviors, and, in fact, most complex behaviors involve many parts of the brain. Leadership and shyness, for example are both known to be very strongly influenced by genes but there is yet little data on exactly which brain regions are involved.

With respect to religiosity there are some clues, though. Patients with temporal lobe epilepsy are often hyper-religious or very preoccupied with religious thoughts, and the evidence is strong that many spiritual leaders and prophets including Mohammed, Joan of Arc, and the apostle Paul suffered from temporal lobe epilepsy. Many other epileptics, even if they are not preoccupied with religion, often experience religious feelings just before they have seizures. Also, experimental stimulation of the temporal lobe in normal non-epileptic people often causes them to feel spiritual sensations. It is also well documented that head injuries can cause a person who is highly religious to become completely indifferent to religion, and can cause a non-religious person to suddenly

WHY ARE PEOPLE RELIGIOUS?

become religious. All of this, of course, indicates that there are, in fact, specific circuits, such as those in the temporal lobe, which generate religious and/or spiritual feelings.

In addition to producing programs and circuits for spiritual feelings there are many ways that genes can encourage or discourage religiosity. Genes, for example, influence logical ability, and, of course, logical ability differs substantially from one person to another. Some people have no trouble accepting that two absolutely contradictory statements are both true. Other people cannot accept logical contradictions; they insist that one or both of two contradictory claims *must* be false. Nonbelievers sometimes argue that the many obvious contradictions in the Bible and the Koran prove that these ancient books cannot therefore be the literal inspired word of God. Others refuse to admit even the most obvious contradictions or don't care.

The Problem of Evil (God is all-knowing, all-powerful, and all-good, but there is terrible evil in the world) has long been cited as a barrier to belief by logical philosophers but it is no problem at all for many believers who don't think as logically or don't require that their religious beliefs are logical.

People differ genetically too in their ability, logic aside, to think rationally and scientifically, in their gullibility, in their inclination to accept authority, and in their fearfulness and their need to believe what is comforting, all of which influence religiosity. Many people, in fact, readily admit that they believe in God and the afterlife more because they want to, more because it is very comforting, than because there is good evidence that these things are true! Other people cannot even imagine believing in something because it is comforting. There are some very big differences in how people think because of differences in genes.

Evolution and Religiosity

Though the evidence is undeniable that genes play a significant role in influencing whether adults are religious or not, some scientists have argued that it is possible that the genes were favored for some other reason, and religiosity is just an accidental byproduct. Other evolutionists argue, though, that it is more parsimonious to assume that the genes were favored by natural selection *because* they made our primitive ancestors religious and thus this enhanced survival. But how so?

Many thinkers believe that our need for meaning is so great that our ancestors, lacking any scientific answers to profound questions about nature of the world and the meaning of human existence, invented gods, and they naturally imagined those gods to be like very powerful people. Belief in supernatural beings, gods and spirits, provided answers and explanations and made existence orderly and meaningful and appealed to the non-rational, emotional parts of the brain, and thus reduced fearfulness and stress and anxiety, and made people feel they were in control, and thus gave them hope, by praying to the gods and appeasing them. A 2010 study published in *Psychological Science* showed that when believers thought about God they had less distress in situations which provoked anxiety. Could these reasons be sufficient to explain why religiosity was originally adaptive? Could the powerful need for meaning by our pre-scientific ancestors have been strong enough that by inventing supernatural answers their fearfulness and anxiety and stress were reduced enough that their survival was enhanced and thus their genes were favored by selection?

Perhaps it was not religiosity per se that was selected for.

WHY ARE PEOPLE RELIGIOUS?

We may not have evolved to believe specifically in gods, but more generally to believe in whatever answers allow us to make sense of our world and to thus feel that we can control our fates and alleviate feelings of helplessness and uncertainty. People have all sorts of irrational beliefs unsupported by science—astrology and superstitions, for example. Our need for meaning may be so great that we are inclined to uncritically accept any system of beliefs that apparently provides answers and the comfortable illusion of control. Our need to make sense of our world and our own behaviors is so basic that when we don't have answers we often make them up. This confabulation is well known to psychologists and it is naturally true of all of us. When amnesiacs, for example, lacking memory of their behavior are asked to explain why they behaved as they did, they invariably make something up, however nonsensical it is. The need to understand our world and especially our own behavior is so primal that we naturally confabulate by inventing explanations. The inclination to believe in religious answers may simply reflect our deep need for meaning.

Nicholas Wade, author of *The Faith Instinct,* has an additional perspective on why religiosity could have been favored by natural selection. In *The New York Times,* November 18, 2009, he said "The ancestral human population of 50,000 years ago, to judge from living hunter-gatherers, would have lived in small egalitarian groups without chiefs or headmen. **Religion served them as an invisible government committing them to put their community's needs ahead of their own self-interest. For fear of divine punishment people followed rules of self-restraint towards members of the community. Religion also emboldened them to give their lives against outsiders. Groups fortified by religious belief would have prevailed** over those that lacked it and genes that prompted the mind toward ritual would eventually have become universal."

Selfish charlatans?

It would, of course, have been personally advantageous for some men in the group to become witch doctors and shamans and priests and other religious leaders, spreading the ideas of gods and supernatural forces and interpreting them in ways to their own selfish advantage. Being apparently able to talk to gods and spirits would have elevated their social status and increased their power and wealth, as is the case today in primitive societies. Even if they were, really, not completely convinced of their own special powers as spiritual leaders, a bit of self-deception would serve them well, as it does to all of us who naturally, subconsciously lie to ourselves. The witch doctors and shamans and priests who encouraged religious belief to their own advantage were not necessarily deliberately deceptive or hypocritical. Self-deception is a normal evolved attribute of the human brain because it has been so critically necessary for us to lie convincingly to one another. The most effective liar is one who believes his own lies.

Belief in gods and spirits and supernatural forces is universal in primitive societies which have little or no knowledge of scientific explanations, and it was probably universal in our ancestors, but that does not mean that all of those people had religiosity genes. Many of them might have been biologically inclined to be skeptical or irreligious, but they had little choice given their environment, except to believe. A couple of hundred years ago virtually all Swedes believed in God. Now only 15% do. This might indicate that Swedes (and generally, most northern and western Europeans) are biologically different from many other world populations in that genes for religiosity are relatively uncommon. Or, it might mean (I think more likely) that biological inclinations toward religiosity can be prevented from expression where societies strongly promote scientific answers and where societies provide adequate social help as in Western Europe and Canada.

Or both explanations could be true. We know that people who are not biologically inclined to be religious often turn out to be non-religious as adults even when they are raised for many years in a religion. It would be interesting to know how often it happens that people who are strongly inclined by their genes to be religious still turn out to be religious even when they are raised by atheists. That would tell us quite a bit about the nature of religiosity genes and how significantly environmental influence can suppress this genetic inclination. And, religiosity aside, it might be very helpful to do behavior-genetics studies of relatives (especially identical twins) raised apart to compare their beliefs in astrology and superstitions and other unscientific ways of explanation. That should offer some insight as to whether the evolved phenomenon we call 'religiosity' is more general than a propensity to believe in gods.

CHAPTER **16**

Why We Age and Die
(And why we won't someday)

PERHAPS THE GREATEST tragedy of human existence is that we are all doomed to die. The most beautiful child, the most loving mother, the most brilliant scientist, the most talented musician, the wisest philosopher; all of us are programmed to deteriorate and die after a few decades because of the process called senescence. One of the greatest mysteries of science is *why* this is so.

Senescence is the predictable genetically-programmed deterioration of organisms which usually begins approximately at the time of reproductive maturity. In humans this means the teen years. Children stay perfectly healthy until about the age of 17 or 18 with no signs of degeneration, but soon after that the deterioration begins. A twenty-five-year-old person has already shown significant signs of senescence, and by the time he is thirty-five the deterioration has progressed enough that he can't compete nearly as well in many physical contests. Forget the Olympics. By the time he is forty his muscular strength has diminished considerably, his skin elasticity and regeneration have been reduced so much he has wrinkles, and his eyesight has deteriorated so much he typically needs glasses.

Mechanical devices often quit functioning because one or more parts wear out or are defective and these weakest links cause the whole system to fail. But that is not what happens with senescence. **All, or most all organs and systems decline in vigor and adaptability and response to challenge at much the same steady rate**. This clearly is evidence of a genetic program. Kidney function, lung capacity, muscle strength, eyesight and response of the immune system all decrease so much that by the time we are sixty or seventy or eighty most of us are so debilitated we succumb to injuries and diseases we would have easily survived when we were younger and stronger.

Sure, a few people beat the odds by a bit, living past 90 and (very rarely) past 100, but nobody has ever lived to be 130 much less 200. (The oldest recorded life span was of a French woman who lived to be 122). The program which makes us degenerate is pretty rigid. No hope that anyone will live to be 130. Not yet anyway. Not until we understand senescence better. Then, maybe we might be able to live forever. More about that later.

Except for a few evolutionary biologists, hardly anyone ever thinks about *why* we die. Why are we programmed to deteriorate? Most of us just accept it as inevitable and assume it is due to wear and tear which can't be avoided. Body parts get old and wear out like machines. That's what most people think.

But that is *not* the case! We have all sorts of repair mechanisms which work perfectly until the senescence program begins when we are teens. Besides, many asexually-reproducing organisms don't senesce at all and some

WHY WE AGE AND DIE

sexually-reproducing species senesce very little if at all. Some tortoises, rockfish, whales, giant clams, and Greenland sharks, for example, have lived for more than two hundred years without any obvious signs of significant degeneration. While annual plants live only one year, sequoias can live for thousands. So senescence is not inevitable for all; constant repair and renewal occurs in some species which get older every year, but do not degenerate, so they are as physically 'young' at 100 as they are at 5.

So the puzzle is why do most sexually-reproducing species senesce at all, and why do some species degenerate tens or thousands of times faster than others? Most mice and small rodents have a maximum lifespan of about three years but some bats can live 30 years. Why do these bats, which are mammals of the same size as mice and rats, live ten times longer? And why do small birds of the same size as small mammals usually live at least twice as long? The oldest known dog lived for 29 years, the oldest cat 36 years, the oldest chimp 59 years and the oldest human 122 years. Why on average do humans live about 40% longer than chimps though we are so closely related. Why do Bowhead whales and Greenland sharks and Galapagos tortoises live 60 or 70 times longer than mice? Why do annual plants live only one year, while biennials live two years while some sequoias live more than 3000 years and some Bristlecone pines can live for 5000 years. Why these enormous differences in life spans?

And why does reproduction seem to be the kiss of death for many species? Mayflies live less than one day after they have reproduced, and Pacific salmon, perfectly healthy for years, live only a few days after spawning! With some species of Australian marsupial mice all the males die a few weeks after mating. Some automatic switch is thrown. Clearly mayflies and salmon and these marsupial mice and many other animals are programmed to die right after they reproduce. Why is this the case?

The association between sex and death is not nearly so immediate with most other species but scientists have found that by delaying reproduction, the life spans of many species can be increased by 50% or more. Women who have never had children, all else equal, generally live longer than women who have reproduced. What is the connection between reproduction and senescence?

Evolution and Senescence

So what are the theories about why we age and therefore die? There are many which focus on proximate causes, damage to DNA, damage caused by mutations, damage caused by free radicals etc. But these are not evolutionary theories. They do not tell us *why*. They don't explain why some species age so slowly or not at all, why life spans differ so radically, why salmon die so soon after spawning but elephants and humans can live more than seventy years after their first reproduction or why there should be a link between reproduction and senescence. Only an evolutionary theory could make any sense out of all these neat patterns and differences.

There is one so-called evolutionary theory which was proposed by Peter Medawar and expanded by the eminent evolutionary biologist George C. Williams. But it is questionable whether it should be called an evolutionary theory because it denies that senescence was favored by natural selection. It basically says that the effects we call aging or senescence are caused by genes which couldn't be removed

WHY WE AGE AND DIE

by selection because these effects happen after the organism has reproduced.

It's true that genes which predisposes youngsters to cancer or heart attacks or anything which kills them before reproductive age cannot be passed on. And the theory correctly points out that genes which cause lethal traits, though, in *old* individuals do get passed on as long as these genes do not cause death until after reproductive age. Such genes slip through the net of natural selection. No way for natural selection to remove them.

One very serious problem with this theory, though, is that it equates the accumulation of deleterious genes with senescence. But senescence is clearly not explained by the random scattershot accumulation of bad genes which happened to slip through the net. If it were, we wouldn't see the enormous variation in life spans with some species living thousands of times longer than others. We wouldn't see progressive predictable deterioration as the animal gets older. We wouldn't see all the neat patterns we see with senescence. Accumulation of bad genes which selection couldn't remove could not produce all the many patterns we see. The evidence, it seems to me, is overwhelming that senescence *had* to be favored by natural selection and the Medawar-Williams theory cannot possibly be correct.

The unique part of the theory that Williams added is the suggestion that *the very same genes* which cause good effects in youth also cause bad effects in old (largely post-reproductive) individuals. He called this negative (or antagonistic) pleiotropy. It's true that many if not most genes are pleiotropic (they produce two or more effects), and certainly some genes have good effects coupled with bad effects, and it's possible that a gene which causes a beneficial effect in youth (like depositing calcium in bones) could, in theory, cause a bad effect later (like causing hardening of the arteries) **but the main problem with this idea is there is no evidence that anything like this causes senescence,** and again it assumes that bad effects are equivalent to senescence.

There are other problems, besides lack of evidence, with Williams's theory. For example why would natural selection favor coupling of good effects with bad ones. Shouldn't selection favor genes which couple good effects with other good effects? Why propose that win-lose situations were favored by selection over win-win? If it's the survival and reproduction of individuals that is important as this theory suggests, it's hard to imagine why natural selection would not favor their continued good health and reproduction indefinitely.

But therein lies the biggest problem with the M-W theory; it assumes that what matters is the survival of individuals. That's what Darwin thought and that's what everyone thought until Bill Hamilton, more than 40 years ago dramatically refined our understanding of natural selection. Hamilton proved that what really matters, what natural selection ultimately favors, is the survival of *genes*. **Individuals are temporary survival machines for genes.**

Sometimes selection favors genes which cause individuals to stop reproducing (menopausal women) or to not reproduce at all, as is the case with scrub-jay helpers–at-the-nest, and worker bees, and naked mole rats. Sometimes selection favors genes which cause their individuals even to *die* in defense of relatives as is the case with parents defending their young and kamikaze defense by worker bees. In the end it's not which individuals survive which matters most—it's which genes. It's often true that the best way genes can reproduce themselves is to sacrifice their individuals, which are after all, only temporary survival machines.

And that's what I think is happening with senescence. The many neat patterns and programs of aging suggest very strongly that senescence did not slip through the net because of lack of selection as the M-W theory suggests, but has been positively favored by natural selection. **In other words the genes which cause us to deteriorate and therefore die have been favored** *because* **they cause senescence.**

But why would genes which cause their survival machines to last only for a limited time do better than genes which programmed their individuals to last indefinitely? Why would genes which cause their model T's to break down and crash at a predictable point in their journey be favored by selection over alleles which caused their vehicles to be immortal? My guess is that this has something to do with vehicular competition. (I am *not* arguing here that individuals die for the good of the species. I am *not* arguing for group selection).

Because of new mutations (some of which are beneficial), and because genes assemble in new combinations (recombination) each generation, and because natural selection is constantly operating, selecting the best new mutations and combinations of cooperative genes, each new generation of organisms is statistically somewhat more fit and superior to the previous generation. After many generations the new models are therefore more fit than the much older generations of their ancestors. So if individuals lasted indefinitely they would be competing with older, less fit models and this competition would reduce the survival of the newer more fit models. Thus genes which got the old models off the streets at an optimal time would do better than genes which permitted the older, less fit models to exist indefinitely. Voila—the evolution of senescence!

What would determine the optimum lifespan for each

species? Why would selection favor very short life spans for some species and very long life spans for others? Why would selection favor life spans in birds that are twice as long as those of same-sized mammals? Why do mayflies and salmon die a few hours or days after reproducing while humans and elephants live for many decades more? Why do some tortoises and clams live a couple hundred years but mice die within three or four years? Could this be explained by vehicular competition? Perhaps some researchers would like to test this. If vehicular competition is, in fact, involved, one would predict that the greater the fecundity (number of offspring of an organism) or at least the greater the number of offspring which survive to reproductive age, the greater the vehicular competition would be and therefore the shorter would be the life span. (Small mammals like mice have many more offspring in a short time than small birds like sparrows thus greater vehicular competition between older and newer generations).

One might predict, as well, that species which are philopatric (offspring live close to parents) would be expected to have shorter life spans than species which are allopatric (offspring disperse farther away from parents) because the vehicular competition would be greater. Could this help explain why clams and tortoises and bats have such relatively long lives?

Another prediction is that animals which provide extensive parental care, like apes and humans and elephants should be expected to live longer, since they are helping the survival of offspring more than animals like rabbits or opossums which provide much less care after birth.

Does it really matter, except to academics, who is right about the reasons why we age and therefore die? Yes, I think it does. The practical and social implications are enormous.

WHY WE AGE AND DIE

If the M-W theory, and others like it which deny that natural selection favored senescence are correct, then there is little or no hope that human life spans can be greatly extended. If we age because of many different deleterious genes which slipped through the net of selection it would seem very unlikely that we will ever be able to counteract their many diverse influences. There would be no possibility of finding any coordinated hierarchy of genes with master controls.

But if, as I think, genes for senescence were positively favored by natural selection *because* they cause senescence, we should expect to find coordinated sets of genes and patterns of mechanisms controlled by master genes just as we do for genes for height or skin color or onset of puberty. Already what we know about progeria (the condition which aging is accelerated so much that children die from senescence by the time they are about 12) suggests there must be master genes which control the rate of senescence. So if the genes which cause us to deteriorate and die were positively favored by selection it should be far easier to prevent expression of these genes, by suppressing the master controls and stopping or greatly slowing the aging mechanisms.

It might be possible that if we avoid fatal diseases and accidents that we could stay young and live for thousands of years—maybe forever. Why not? Senescence is not inevitable. We are not like machines which eventually wear out. We age because of genes which program us to age and therefore die. We can very likely block those genes, so we remain indefinitely with bodies like those of twenty-year olds.

This would, of course, cause the greatest social and political and ecological changes ever in human history. Someone might want to write a novel with the premise that we, in the future, will live for thousands of years if not forever.

◀ **CHOOSY WOMEN AND CHEATING MEN**

Would it be a good thing? The problems would be staggeringly enormous. My guess is, though, if they are selling the pills at Walgreens, almost everyone will buy them. I don't think many of us will choose to die for the good of the species.

CHAPTER **17**

Evolution In The Third Grade

Intelligent life on a planet comes of age when it first works out the reason for its own existence. If superior creatures from space ever visit earth, the first question they will ask in order to assess the level of our civilization is: 'Have they discovered evolution yet?' —Richard Dawkins

IN PUBLIC SCHOOLS in the United States, Canada, Western Europe, Australia, and most of Asia and South America, evolution is commonly taught but not until students take high school biology classes. Ideally, though, when should students first start learning about evolution? Isn't learning about evolution so extremely important that instruction should begin much earlier than high school? Perhaps as soon as the third grade?

There are many countries where students do not learn anything at all about evolution. In many African countries, in fact, conditions are so primitive that few opportunities exist for formal education even as far as high school. It is often such a struggle to survive to adulthood that few youngsters learn much about science, much less about evolution.

In many other countries public schools provide education for most youngsters through high school but there is strong opposition to teaching evolution, especially human evolution.

This is true of most Muslim-majority countries. In Pakistan evolution is rarely taught even at the university level. Likewise, most students in Turkey and Indonesia and Malaysia never learn anything about evolution. There are, of course, many school districts in several U.S. states where evolution is not included in high school biology classes. Many of them are in the Bible Belt of the south. Christian fundamentalists like Islamic fundamentalists instead believe in intelligent design and creationism, and reject scientific explanations of the evolution of life. They believe the Koran and the Bible, not the scientists.

So what do we know about evolution in the 21st century which would suggest that all children in public schools should learn about evolution, the earlier the better?

More than 150 years ago, Charles Darwin (and Alfred Wallace) after many years of superb research, proved to the scientific world that evolution is a fact. They proved that life on earth has evolved by a natural process and can be explained without reference to supernatural forces. Since then, the evidence for evolution has grown far stronger, and there is, after all this time, still no scientific evidence which contradicts their awesome discovery. The fossil evidence and the biogeographic evidence they relied upon, for example, is today hundreds of times greater than it was then, and *all* of it confirms that Darwin and Wallace were correct.

Also, today we have powerful genetic evidence which corroborates what is known from the fossil record and biogeography. All the more amazing that Darwin and Wallace were able to figure it out with only a tiny fraction of the evidence

EVOLUTION IN THE THIRD GRADE

we have today. They had no knowledge of DNA, or genes, or the enormous body of confirming evidence for evolution we have today from embryology (such as gill slits and fish-type kidneys and tails in human embryos). Nor did they know about the enormous evidence we have today from dozens of modern scientific fields which didn't yet exist in their day.

Darwin's publication of *The Origin of Species* in 1859 started what many historians and philosophers of science maintain is the greatest revolution in human thought of all time. Biology was dramatically revolutionized and made much more powerful as evolution became central to all biological explanation. As Dobzhansky said, "Nothing in biology makes sense except in the light of evolution."

It was not just biology which became radically changed and more powerfully explanatory; much of classic philosophy and many traditional religious beliefs were perceived to be antiquated and irrelevant and obviously wrong to the increasing numbers of scientific thinkers who accepted evolution. The revolution had begun.

As indicated in chapter 2, Darwin's most impressive contribution was not, perhaps, the discovery of the fact of evolution, but his realization of the principle of natural selection, the process by which most evolution happens. Darwin showed us how to get past proximate, 'who', 'what', 'where', and 'when' answers, and for the first time in human thought how to scientifically and ultimately answer 'why' questions about life. Why do some people have dark skin and others are light? Why do males of most species court females and fight more, and why are they more promiscuous than females? Why do birds sing? Why do desert plants have spines? Why do humans and all animals have leftover useless vestiges, and why is biological design so poor? Why are biological traits this way and not that? Millions of 'why' questions about life could now be scientifically and ultimately answered. **Darwin showed us how to understand the meanings of life.**

The evolution revolution fueled by understanding of natural selection continued to steadily grow after Darwin. The best universities in the U.S. and throughout Europe made evolutionary science central to the study of their biological science programs, and high schools increasingly included units on evolution in their biology classes. The study of biology was dramatically changed forever.

And so eventually was Philosophy changed so radically that it is questionable nowadays whether a Philosophy major could be considered competent without a solid understanding of evolutionary explanations. Sure, if you want to be well educated today in the liberal arts and sciences it is useful to learn a bit about the ancient philosophers like Plato and Aristotle and the many classic philosophers since, such as Hume and James and Descartes. It is helpful to take a History of Philosophy class and learn what those philosophers thought then, and how they thought we should think, about who we are, and how to know what is true and good and moral. And, you might win some money on Jeopardy.

But if you are impatient and want to get right to the meaning of life; who we ultimately are, and why everything biological is the way it is, and even if you wish to understand the origins of ethical and moral behavior, you might want to follow the advice of the eminent biologist George Gaylord Simpson who suggested, in effect, that you could skip over all the philosophers before 1859. The golden age of philosophy began with Darwin.

The Evolution Revolution continued to steadily grow in the 21st century dramatically improving our understanding of the nature of life and the meaning of human existence, and it was not just the biological sciences and Philosophy which were changed. Even some religions adapted and modernized; the Catholic Church, for example ever since 1996 at least, when John Paul II was Pope, has officially accepted evolution and has proclaimed that there is no conflict with its teachings. The Church, though, has maintained that humans were given a soul by God, and that evolution has been purposely planned. Virtually all evolutionary scientists, however, disagree; they insist that the evolutionary process clearly is not planned and has no purpose. An excellent discussion of this history can be read by consulting Wikipedia on "The Catholic Church and Evolution".

How Hamilton Changed Evolutionary Science

It turns out that 100 years after Darwin the Evolution Revolution was still in its infancy. In the 1960's it got a powerful growth shot in the arm with the publication of William Hamilton's *The Genetical Evolution of Social Behaviour*. Influenced by the writings of previous scientists Hamilton radically refined and improved our understanding of natural selection by demonstrating that the essential unit selection operates upon is not the individual as Darwin thought, but the gene! (Selection operates upon replicators; genes are replicators, individuals are not). As pointed out in the chapter *Charity Begins at Home*, Darwin did not know about genes and so there were many problems which Darwin realized raised apparent contradictions to his theory of natural selection. How, for example, to explain altruistic behaviors and the sterile castes of social insects. If selection favors the fittest individuals how could it favor altruism, even self-sacrifice for others, and how could it produce individuals like worker bees which don't ever reproduce.

Hamilton's gene-focused refinement of natural selection proved that Darwin had nothing to worry about; there were no contradictions. Altruism and sacrificial behavior and sterile castes could all be explained once one understood that 'survival of the fittest' should refer to genes, not individuals. Individuals, by this view, are temporary survival machines for genes. **Almost overnight natural selection theory became far more powerfully explanatory and predictive and far more applicable to understanding behavior.**

By the 1970's E.O. Wilson of Harvard, applying this new gene-focused understanding of natural selection, published his awesome synthesis *Sociobiology* (perhaps the most important scientific book of the century) establishing a great new science and the most scientifically convincing basis ever for understanding the social behaviors of animals (and humans too), though at first some social scientists were outraged that Wilson dared to say that the same basic principles which explain social behaviors of animals are also applicable to understanding human behavior.

Soon Richard Dawkins's *The Selfish Gene* and many other popular books describing gene-focused selection theory were being read by millions of people who for the first time could understand why animal and human behaviors evolved to be the way they are. Many other academic and popular books written by leading scientists and scholars since the 80's until the present have contributed as well to our rapidly growing understanding of behavior and many university departments added "Evolution" and "Evolutionary" to the names of their programs in the biological sciences, psychology and anthropology.

The field of Evolutionary Psychology got it's start in the 70's and has been growing rapidly ever since, and is now so central to the study of human behavior at many colleges and universities that students in the behavioral sciences cannot hope to understand much about human behavior if they do not have a good understanding of this discipline. Wikipedia, again, has an excellent and thorough discussion of the history and present status of Evolutionary Psychology, and there are several texts and many books which describe this field and its applications, from altruism to mate choice and gender behaviors. Introductory psychology courses today which do not include much discussion of Evolutionary Psychology are almost as lamentably inadequate as introductory biology courses with little or no discussion of evolution.

Our improved gene-focused understanding of natural selection is today being used to provide insight in many fields which until recently were poorly informed or unfamiliar with applications of evolutionary science. Take medicine for example. Though it has long been recognized that understanding of natural selection is critical for immunology, the development of effective vaccines and the treatment of many medical conditions, the new science of Darwinian medicine, which applies modern selection theory to understanding diseases and their treatment, promises to greatly improve medical science in the future. Williams's and Nesse's 1994 book *Why We Get Sick*, beautifully written for the general audience, describes how diseases and symptoms must be viewed through a modern evolutionary lens if good medical treatment is to be achieved. Are the symptoms of an illness evolved adaptations whereby the pathogen is manipulating the behavior of the patient causing coughing or diarrhea, for example, to spread it's own genes by infecting other hosts, or are these symptoms adaptations which have evolved as defenses against the pathogen like some forms of coughing or diarrhea? It makes a big difference how and whether the symptoms should be treated.

Fever caused by bacterial infections has evolved as a natural body defense. Raising the body temperature helps the patient recover by interfering with the bacteria. Depending upon the age and condition of the patient it can thus be a serious mistake to reduce the fever by giving aspirin or acetaminophen. At the very least this is likely to prolong the illness and at the worst in some cases it may risk the life of the patient. If the fever is so high as to cause seizures or other damage it may be best to reduce it, but evolutionary medicine would suggest it would not be wise to give aspirin or acetaminophen to a toddler with a bacteria-caused temperature of 102. On the back cover of *Why We Get Sick,* Michael Gazzaniga says "This is the most important book written about issues in biomedicine in the last fifty years" and Richard Dawkins says "Buy two copies and give one to your doctor."

Perhaps the most important impact of the modern revolution in evolutionary science has been on the study of human behavior, the main focus of this book. In the past three decades there has been dramatic new scientific understanding of gender differences, sexual behaviors, altruism and cooperation, and selfishness, human aggression and homicide, political behavior, team sports, patriotism and warfare, evolved emotional responses and motivations, natural attractions and aversions, aesthetics, cognition, learning, and many other human behavior topics. All of these have been made far more understandable today by the evolutionary psychology perspective that our behaviors were selected because they were adaptive for ancestral environments often very different from those we experience today. Thus to understand our common behaviors shared by all human societies, it is necessary to identify the ancient environmental circumstances under which our behaviors evolved. That's what Evolutionary Psychology does and that's why it is revolutionizing our understanding of ourselves.

And that's why if you are a student today intending to major in any of the behavioral sciences, especially Psychology and Anthropology, or if you wish to major in Philosophy or to pursue a career where it is important to understand human behavior, you had best pick a college with a strong emphasis on Sociobiology and/or Evolutionary Psychology. Among the best are Harvard, University of California at Santa Barbara, University of Texas at Austin, University of Michigan at Ann Arbor (with it's Evolution and Human Adaptation program), the University of New Mexico, and there are many others: Stanford University, University of Pennsylvania, University of Oregon, University of Miami etc. In Canada, one of the best places is McMaster University in Hamilton, Ontario thanks to Evolutionary Psychology leaders Margo Daly and Martin Wilson. If you are not particularly interested in human behavior but want to be a general biology major check to make sure that the Biology Department not only has a strong emphasis on evolutionary science but that their curriculum reflects *modern* evolutionary science and is not stuck with a pre-1960s interpretation of natural selection theory.

The Third Grade?

So when should children first start learning about evolution? Isn't evolution the most important scientific discovery of all time? Didn't Darwin show us how to scientifically answer 'why" questions about life? Doesn't evolutionary science provide the best explanatory and predictive framework ever conceived for understanding the living world? What could be more necessary to a basic education than learning about evolution? Shouldn't we be teaching students the basics about evolution and natural selection much sooner than waiting until high school?

Well, the British government thinks so. Legislation has recently passed which makes evolution education compulsory as of September 2011 in the primary schools of England. The new curriculum includes the requirement to explain how plants and animals adapt to their environment by natural selection. Even church schools within the state system will be required to educate their students about evolution though they will be permitted considerable flexibility as to how they do this. Professor Sir Martin Taylor, Vice President, of the Royal Society said "We are delighted to see evolution explicitly included in the primary curriculum. One of the most remarkable achievements of science over the past 200 years has been to show how humans and organisms on the Earth arose through evolution."

CPSIA information can be obtained at www.ICGtesting.com
Printed in the USA
BVOW031415040213

312355BV00001B/13/P